The Renaissance of Remote Places

With a particular attention to remote places and marginalized territories, this book provides a conceptualization of the role of internal and international migration to the local development and resilience of the rural and mountain regions of Europe.

The book is a collective effort produced by the international and multidisciplinary network of the Horizon 2020 project MATILDE. In declaring a public and transregional position – in the form of a Manifesto for the renaissance of remote places – the book contributes to a new narrative about migration and rural/mountain territories for the future of the entire continent. Mobilizing new data and scientific-based information, the book calls for putting remote regions and their inhabitants at the core of innovative policies at local, regional, national and EU levels.

An important resource for researchers, students and policymakers in human and population geography, rural studies, migration studies, social and political sciences.

Andrea Membretti is Assistant Professor of Sociology at the University of Pavia (Italy) and is currently Senior Researcher at the University of Eastern Finland (UEF).

Thomas Dax is Senior Researcher on Rural and Mountain Development Studies, Federal Institute of Agricultural Economics, Rural and Mountain Research, Vienna, Austria.

Anna Krasteva is Professor of Political Sciences, President of Policy and Citizens' Observatory, Doctor Honoris Causa of University of Lille, France.

Routledge Studies on Remote Places and Remoteness
Series editor: Andrea Membretti

The Renaissance of Remote Places
MATILDE Manifesto
Edited by Andrea Membretti, Thomas Dax and Anna Krasteva

For more information see our webpage: https://www.routledge.com/Routledge-Studies-on-Remote-Places-and-Remoteness/book-series/RSRPR

The Renaissance of Remote Places

MATILDE Manifesto

Edited by Andrea Membretti, Thomas Dax and Anna Krasteva

MATILDE Editorial Manager: Mia Scotti

First published 2022
by Routledge
4 Park Square, Milton Park, Abingdon, Oxon OX14 4RN

and by Routledge
605 Third Avenue, New York, NY 10158

Routledge is an imprint of the Taylor & Francis Group, an informa business

British Library Cataloguing-in-Publication Data
A catalogue record for this book is available from the British Library

Library of Congress Cataloging-in-Publication Data
A catalog record for this book has been requested

ISBN: 978-1-032-19711-1 (hbk)
ISBN: 978-1-032-19712-8 (pbk)
ISBN: 978-1-003-26048-6 (ebk)

DOI: 10.4324/9781003260486

Typeset in Times New Roman
by Apex CoVantage, LLC

The cost of publishing this book was covered by the MATILDE project. MATILDE has received funding from the European Union's Horizon 2020 research and innovation programme under grant agreement No 870831.

The European Commission is not responsible for the content of this publication.

Contents

Illustrations

Figures

Table

Contributors

Birgit Aigner-Walder is Professor for Economics at the Carinthia University of Applied Sciences (Austria) and Head of the Department Demographic Change and Regional Development of the Institute for Applied Research on Ageing. She does research on the economic effects of demographic changes with special focus on the labour market, consumption and services of general interest.

Simone Baglioni holds a chair in sociology in the Department of Economics and Management at the University of Parma, Italy. His work focuses on issues of migration and labour market, civil society and social innovation and platform capitalism.

Filippo Barbera is Professor of Economic Sociology at the CPS Department of the University of Turin and Fellow at the Collegio Carlo Alberto (Torino). His research interests are social innovation, foundational economy and development of marginal areas. His recent publications include *Alternative Food Networks: An Interdisciplinary Assessment*, 2018, Palgrave Macmillan (eds. with A. Corsi, E. Dansero, C. Peano), and *The Foundational Economy and Citizenship*, Bristol, Policy Press 2021 (eds. With I.R. Jones).

Fabrizio Barca, Statistician and Economist, Co-coordinator of the Think-and-Do Tank Forum on Inequality and Diversity, has been Minister for territorial cohesion, Chief at the Italian Treasury, President of the OECD Territorial Policies Committee, Special Advisor to EU commissioners, Research Chief at the Bank of Italy. He has taught in EU universities and written on corporate governance, capitalism, policy-making. He is a member of the MATILDE advisory board.

Maria Luisa Caputo is a postdoc researcher at the University of Parma for the MATILDE project, since July 2020. She holds a PhD from the University of Paris 1 Panthéon-Sorbonne. She previously worked for two years as a research and teaching attachée (ATER) at the University of Paris Est-Créteil. Other previous experiences include working as a research assistant at COMPAS, University of Oxford, for the INTERACT project on TCNs' integration as a three-way process, and at CNRS, France, for the SIESTA project on the territorial dimension of the EUROPE 2020 Strategy.

Cristina Dalla Torre is a researcher at Eurac Research – Institute for Regional Development (Italy) and PhD fellow at University of Padova – Department of Land, Environment, Agriculture and Forestry (Italy). She has been researching on social innovation and on governance of collective natural and built resources in mountain areas using a transdisciplinary approach since 2017.

Thomas Dax, Federal Institute of Agricultural Economics, Rural and Mountain Research, PhD in Social and Economic Sciences, Senior Researcher on Rural and Mountain Development Studies. He is promoting research cooperation on rural change and mountain policy assessment at the international scale and is a member of numerous international working groups, including a longtime participation in the OECD Working Party on Rural Policy and the Working Party on Territorial Indicators.

Marika Gruber is Senior Researcher and Lecturer for Inter- and Transculturalism at the Carinthia University of Applied Sciences; Deputy Head of Department Demographic Change and Regional Development (DCRD) – Institute for Applied Research on Ageing (IARA); Member of the research group Trans_Space – Transformative Societal, Political and Cultural Engagement. Her research focus is on migration studies and rural areas, social and labour market inclusion, integration governance, participatory and transdisciplinary research.

Ulf Hansson is a lecturer (PhD) in Political Science at Dalarna University and a researcher in the MATILDE project. His research interests range from the role of education in divided societies to issues surrounding migration and integration in a rural context.

Ayhan Kaya, Professor of Politics, Istanbul Bilgi University; ERC AdG Holder; Jean Monnet Chair of European Politics of Interculturalism and Member of Science Academy, Turkey. He received his MA and PhD from University of Warwick. He has specialized on migration, youth radicalization, European integration, diasporas, refugees, populism, Islamophobism, Islamism.

Stefan Kordel holds a PhD in Geography and passed a habilitation during his work as a postdoctoral researcher at the Institute of Geography at the

University of Erlangen, Nuremberg, Germany. His research focuses on various processes of immigration to rural areas and its implications for rural development as well as sociogeographical questions around place-based belonging and attachments in transnational settings. Furthermore, he cochairs the working group for rural studies within the German Society of Geography.

Anna Krasteva is Professor of Political Sciences; Doctor Honoris Causa of University of Lille, France; Director of CERMES (Centre for European Refugees, Migration and Ethnic Studies) at the New Bulgarian University; President of Policy and Citizens' Observatory: Migration. Digitalization, Climate, Editor-in-Chief of the international journal Southeastern Europe (Brill).

Jussi P. Laine is Associate Professor of Multidisciplinary Border Studies at the Karelian Institute of the University of Eastern Finland and holds the title of Docent of Human Geography at the University of Oulu, Finland. Currently, he also serves as President of the Western Social Science Association (WSSA) and in the Steering Committee of the IGU's Commission on Political Geography.

Raúl Lardiés-Bosque, Department of Geography and Regional Planning, University of Zaragoza, Spain. PhD on Geography. His main research interests focus on population geography, and specifically on migration and residential mobility, tourism geography and ageing in relation to quality of life.

Per Olav Lund, Researcher. Specialization in analyses of regional development and public services. Participant and Project Coordinator for the Norwegian MATILDE team. Main research interests are in municipality economics, social and economic welfare and impacts.

Dr. Ingrid Machold (F) graduated from the University of Vienna with a PhD in Sociology. She is Senior Researcher at the Federal Institute of Agricultural Economics, Rural and Mountain Research and Deputy Head of the Department Mountain Areas Research and Regional Development.

Andrea Membretti, PhD in Sociology, is currently Senior Researcher at the University of Eastern Finland (UEF), Karelian Institute. He is Assistant Professor of Sociology of the Territory at the University of Pavia and Faculty Member in Social Sciences at GSSI – Gran Sasso Science Institute (Italy). He is also Research Fellow at the University of the Free State (South Africa) and Affiliate Researcher at the Department of Culture, Politics and Society of the University of Turin (Italy). His main field of study is migration to and from mountain, rural and remote regions, in relation to socioeconomic and demographic transformations, climate change and extreme events. He has been Scientific Coordinator of the H2020 MATILDE Project.

Nuria del Olmo-Vicén Department of Sociology and Psychology, University of Zaragoza, Spain. PhD on Social and Political Sciences (EUI, Florence). Her main research interests focus on social processes in multicultural societies, and particularly on the impact of religious and cultural factors in the collective identity construction among immigrants.

Manfred Perlik is a professor associated at the Centre for Development and Environment at the University of Bern, Switzerland, and also at UMR Laboratoire Pacte at the Université Grenoble-Alpes, France. As an economic geographer, he focuses on urbanization of mountain areas, including processes of migration and multilocality, as well as questions of spatial justice and transformative social innovation.

Rahel M. Schomaker is a professor for Economics and Public Administration at Carinthia University of Applied Sciences, and WSB, an adjunct professor at the German University of Administrative Sciences and a senior fellow at the German Research Institute for Public Administration and the IARA. Her research focuses on administrative change, crisis governance, migration and trust.

Susanne Stenbacka is a professor in Human Geography at Uppsala University. Her research revolves around rural transformation, regional development and rural-urban relations, with specific focus on gendered identities, migration and provision of welfare services. Her current studies deal with refugee migration to rural areas, centre-periphery relations in the aftermath of police withdrawal and rural experiences of disability.

Dr. Tobias Weidinger works as a postdoctoral scholar at the Institute of Geography at the University of Erlangen-Nuremberg, Germany. He holds a PhD from the same university and is interested in diverse processes of (im)mobilities in rural areas and the development of qualitative and participatory space-related methodology.

Annelies (E.B.) Zoomersis Professor of International Development Studies (IDS) at Utrecht University and Chair of WOTRO Science for Global Development (NWO). She is the cofounder of the Netherlands Land Academy (LANDac) and the founding chair of Shared Value Foundation (SVF). Annelies Zoomers is currently the PI of Welcoming Spaces – revitalizing shrinking areas by hosting non-EU migrants. This programme is carried out with partners in the Netherlands, Germany, Spain, Italy and Poland (www.welcomingspaces.eu/). She is a member of the MATILDE advisory board.

1

Introduction

The renaissance of rural, mountainous and remote regions of Europe

A call for action

Andrea Membretti, Thomas Dax and Anna Krasteva

Rediscovering the rural "backbone" of the European Union

The regions – as institutions in between the state and local governments – have been assuming a leading role in the process of European integration in the recent past, being considered in many respects the veritable "backbone" of the European Union. Towards the turn of the century, the 1990s saw numerous important developments and a number of optimistic predictions concerning improvements in territorial balance and EU governance (Magone, 2003). During that period, there was an evident shift to the *Europe of the Regions* perspective, in which the main avenues of the EU policy intended to overcome territorial inequalities favoured the active role of these *intermediate bodies*, implying an increased need for the so-called *multilevel governance approach* (Charbit, 2020).

However, the 2000s witnessed a progressive reduction in this respect as European institutions paid increasingly less attention to these territorial actors, at least in regard to their role in a participatory and inclusive governance of the EU. This happened despite the heightened role of the regions in the construction of the European Union, as well as in the design and implementation of its policies, and also despite the debate on territorial cohesion promoted by the publication of the European Commission's Green Book: "Territorial Agenda process and discussions on Multi-level governance" (EC, 2008; BMI, 2020) (Commission of the European Communities, 2008). Even though various funds were invested in the local development of these regions and substantial financial means of regional policy were deployed in order to alleviate territorial imbalances or enhance regional performance (by allocating up to one-third of the EU's budget to these measures), the territorial effects remained limited. Therefore, in recent decades, the sensation of being on the margins of economic and social policies has grown stronger. This is particularly true of rural, remote

DOI: 10.4324/9781003260486-2

and mountainous regions: many of them have long considered themselves to be *places left behind* or *places that don't matter*. It is no coincidence that in these territories, disaffection with European institutions is spreading, populism is growing and xenophobic forces or sovereignist movements are emerging (Rodriguez Pose, 2018).

However, it should not be forgotten that half of the European land mass is classified as predominantly rural, and about 30% of it as mountainous. Furthermore, the marginalization of rural and mountain regions is particularly objectionable if we consider Article 174 of the Treaty on the Functioning of the European Union, which states that the EU shall strengthen economic, social and territorial cohesion within the EU, in particular by "reducing disparities between the levels of development of the various regions and the backwardness of the least favoured regions." It goes on to state that:

> Among the regions concerned, particular attention shall be paid to rural areas, areas affected by industrial transition and regions suffering from severe and permanent natural or demographic handicaps, such as the northernmost regions with very low population density and island, cross-border and mountain regions.

Despite their long-standing neglect, the core role that these regions can play for Europe's shared wealth and well-being is clear for all to see. The agricultural production, forests, water reserves, cultural heritage, bio- and social diversity, languages and local autonomy of these areas make them simply irreplaceable.

Therefore, it seems particularly important that, after decades characterized by a strong focus on urban and metropolitan territories (and on spatial and socioeconomic agglomeration processes), a shift of attention towards these territories has finally been shown by European institutions, together with a new awareness among public opinion. With the launching of the *long-term vision for the EU's rural areas*[1] in 2021, the European Commission in fact recognizes that:

> A vibrant tapestry of life and landscapes, Europe's rural areas provide us with our food, homes, jobs, and essential ecosystems services. To ensure that rural areas can continue to play these essential roles, a European Commission communication sets out a long-term vision for the EU's rural areas up to 2040. It identifies areas of action towards stronger, connected, resilient and prosperous rural areas and communities. A Rural Pact and an EU Rural Action Plan with tangible flagship projects and new tools will help achieve the goals of this vision.
>
> (EC, 2021)

In this unprecedented context, the president of the European Commission, Ursula von der Leyen, has declared that "rural areas are the fabric of our society and the heartbeat of our economy. They are a core part of our identity and our economic potential. We will cherish and preserve our rural areas and invest in their future."

Furthermore, in the face of the radical changes imposed by the COVID-19 pandemic, what these regions have to offer in terms of differing modes of settlement, production and consumption is likely to be increasingly sought after, because their local systems are characterized by less anthropic pressure and are better able to shift towards circular economies. Because densely populated urban areas have been hit hardest by the pandemic, the social rarefaction of these regions connotes welcome resilience amid this time of crisis.

It therefore seems that, despite the long period in which the regions – these fundamental "intermediate bodies" at the very core of EU governance – have received little political attention, global phenomena with a high socioeconomic impact, such as the climate crisis and the pandemic, are highlighting the enormous value of local autonomy and administrative decentralization, in particular in regard to mountain and rural areas.

Immigration in the places left behind: a resource for local resilience and regional development

The MATILDE Manifesto starts from the momentum visible in these recent shifts: from the potential of these *places left behind*, and from their desirable and feasible renaissance. It considers the fundamental contribution that newcomers – together with locals – can make to this process.

When analysing the potential of these regions, immigration (and "new peopling" more in general) is one of the fundamental factors that must be considered. Rural, mountain and remote areas have been structurally losing inhabitants for decades: they suffer from chronic labour shortages often due to the flight of young people to the cities. Consequently, they are increasingly suffering from ageing populations. Naturally, the arrival of new inhabitants – both internal and international migrants – is a fundamental resource. As widely demonstrated by the data collected by the MATILDE Project (Laine, 2021; Caputo et al., 2021), intra-EU migration, interregional immigration, as well as international immigration from non-European countries, today make the main contributions to the demographic stability of marginalized regions across Europe, as well as being factors central to the functioning of entire sectors of local economies, from agriculture to tourism, from personal services to small- and medium-sized industrial or craft firms.

It therefore seems crucial to discuss, on the basis of scientific results, the extent to which migration weighs on the development of rural, remote and mountain regions, by not only focusing on the fundamental integration processes of foreigners or on the identification of policies in favour of newcomers, but also by exploring migration's impact on socioeconomic and spatial changes, considering how it affects the overall development of these regions.

When considering migration, one must acknowledge and address also its challenges. Local populations and immigrants will never be able to establish an inspirational vision for their common future in these regions if policies are not developed for the proactive inclusion of new inhabitants, creating new citizens; if public spaces of encounter and negotiation between cultures and needs are not established at the very local level and if rural and mountain communities are not, to a certain extent, entitled to preserve their own traditions compared to those of outsiders, within well-governed and long-term processes in which new cultural syntheses and innovations may be finally achieved (Membretti et al., 2017).

Migration research (in particular when it is action research, as in MATILDE) can give the marginalized regions of the continent a unique occasion to develop participatory reflection on their capacity for resilience, on their creative adaptation to current and future challenges, and doing so while the relationships between rural/mountain and urban dimensions, between the local and global scales, are rethought.

The MATILDE Project: participatory action research and local engagement

Founded by the European Union in the framework of the Horizon 2020 programme, MATILDE (*Migration Impact Assessment to Enhance Integration and Local Development in European Rural and Mountain Regions*) is a three-year research project aimed at questioning, rethinking and reconceptualizing the nexus between migration and local development by examining how foreign immigration impacts on socioeconomic well-being and territorial cohesion in European marginalized and remote regions.

The project is the result of previous research and public reflection on the topic of migration in mountain areas (Perlik et al., 2019; Membretti et al., 2018) carried out since 2015 by the international *ForAlps* network (www.foralps.eu). Under the coordination of the University of Eastern Finland (Karelian Institute), a large consortium of 25 academic partners and local organizations (NGOs, associations, provincial administrations, etc.) has been set up and is carrying out research in 10 European countries, involving some 15 case studies at regional and local level.

The MATILDE Project's basic assumption is that, if unaddressed, the sentiments of people living in *places that don't matter* risk fuelling an authoritarian dynamic, rejecting diversity *a priori* and offering electoral support for antielite and xenophobic parties (Barca, 2019): immigrants in rural areas therefore risk being considered an additional burden, or even a threat to local inhabitants, and consequently treated as scapegoats. In fact, people living in marginalized areas have been experiencing in recent decades the dramatic shift of both private and public services to major cities that often gives rise to a striking demographic decline and an ageing population (Copus et al., 2021). While also gaining a role as destinations of international migration, the specific needs of rural and mountain regions are not adequately reflected in the governance of migration, either at national or regional levels.

To understand the perceptions, misperceptions and perspectives of local actors comprehensively, the MATILDE Project's exploration of local-level interaction adopts a participatory action research approach (Lewin, 1946), which means conducting research *with* rather than *on* the subjects of study. Accordingly, the MATILDE case study regions (ranging from Scandinavia to the Balkans and Anatolia, as shown in the map) enable a process of knowledge coproduction between researchers and participants that brings about a change in perceptions and practices. The case study areas have been selected in order to guarantee representativity of spatial and historic characteristics in terms of migration patterns and governance, welfare systems, sociocultural and economic systems. Moreover, the selection represents the heterogeneity of rural and mountain areas in terms of the degree of urbanization, remoteness and population density, geographical characteristics and specific migration-related structures.

In particular, the MATILDE Project examines how the distinctive features of rural and mountain contexts interact with migrants' integration paths and impacts, considering that demographic trends, socioeconomic dynamics and migration patterns in a specific region affect opportunities, policy responses, societal attitudes and perceptions of newcomers.

Adopting this perspective, developed within the project has been an innovative conceptual and methodological framework (Kordel and Membretti, 2020) that promotes a change of perception and practices through the identification of spatially located and path-dependent factors.

In the midst of its research and public awareness activities, the MATILDE Project has already collected a large amount of quantitative and qualitative data (all available at www.matilde-migration.eu), and it has launched a series of extensive participatory processes in all the regions involved. The *Manifesto* presented here constitutes one of the main activities of collective reflection and public communication in which the participants in the project have collectively engaged.

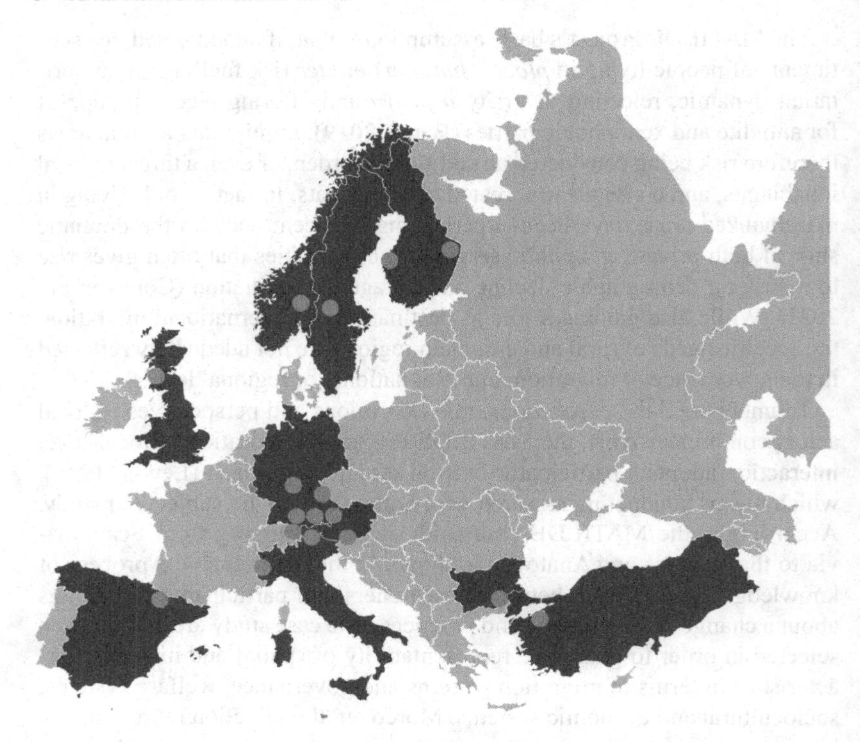

Figure 1.1 MATILDE countries and case study regions
Source: MATILDE Project

From misperceptions to recognition: new data, new policies

When considering the actual and potential role of newcomers in the development of marginalized regions, it should be acknowledged that European public opinion significantly overestimates the number of non-EU immigrants: in fact, in 19 EU member-states, citizens perceive the proportion of immigrants as at least twice the actual figure (Eurobarometer, p. 469). Particularly after the 2015/2016 peak in refugee arrivals, the topic gained central stage in the political debate, with multiple consequences. Indisputably, misperceptions regarding the impact of international migration have led to the polarization of politics in EU member-states and to the rise of populist and even racist political forces across Europe (Inglehart and Norris, 2016). At the EU level, this has given rise to the unprecedented and ongoing political crisis caused by the failure to reach agreement on the distribution of asylum seekers. Hostility towards migration also negatively affects citizens' attitudes to social redistribution in general (Alesina et al., 2018). With significant variations among countries, attention is centred on how to limit the extent and impact of migration, rather than on constructing a governance system with which to untap the development potential of migration in the countries and regions of destination (Perlik and Membretti, 2018).

Wide misperception also affects the spatial distribution of international migrants in Europe and their actual contribution to the hosting societies and national/regional economies. Against the background of global trends such as urbanization and agglomeration, it should be recognized that economic and forced migration flows are also increasingly oriented to locations outside urban areas. This is the result of the free movement of economic migrants and the effects of dispersal policies[2] targeting asylum seekers and refugees. Recent studies on the features and implications of immigration to rural Europe highlight the development potential for rural and mountain areas (Kordel and Membretti, 2020; Galera et al., 2018; Membretti et al., 2017). Migration plays a key role in demographic processes across Europe, even more so in rural and mountain regions experiencing demographic decline. While migrants' employment in these areas is marked by seasonality, geographic concentration and ethnic labour niches, demographic changes and the development of nonagricultural activities open new opportunities. Among them, the increasing demand for services related to an ageing population, in the tourism and construction sectors, in food production and distribution has the highest potential for employment effects (Bianchi et al., 2021).

Noticeable at the same time is the substantial lack of scientific knowledge and reliable data regarding the development potential and impact that international migration brings to rural and mountain regions. With migration studies still mostly centred on urban areas, there is a lack of conceptualization about this new phenomenon as well. Notwithstanding the prominence of urbanization as a global trend (at least until the current pandemic), it is time to analyse the role that migrations to rural and mountain areas can play for European rural and even remote regions, among other things, by contributing to the revitalization of social and economic local milieus, reducing territorial inequalities and taking part in urban-rural interconnections.

It is also time to investigate and to open a public debate on how immigration can increase diversity in marginalized territories of Europe and create opportunities for social innovation stemming from social rarefaction rather than agglomeration (Remotti, 2011), and how it can be a crucial factor in attaining balanced territorial development, as defined in the UN Sustainable Development Goals (Target 11.a, "Strengthening development planning").

To avoid the risk that immigration flows have a negative impact on socio-economically and geographically fragile areas, place-sensitive policies and adequate governance measures are needed, also considering that local development programmes for rural regions have long been in place (like LEADER/CLLD). Importantly, their remit should more explicitly incorporate the potential to use social innovation processes and focus on the beneficial role of social diversity for the local development of rural and mountain regions and the achievement of spatial justice (Dax et al., 2016; Shucksmith et al., 2020).

To date, dispersal policies for the reception of asylum seekers and refugees have been experienced by rural areas as additional burdens on already-marginalized territories. At the same time, the rural employment of economic migrants has been often regarded as merely a match between unskilled workers and jobs that natives no longer want (hiding the fact that quite often the skills and educational qualifications of migrants are not recognized and valued in their new living places).

Why a *Manifesto*? The aim and structure of this book

The MATILDE Project has the aim of producing scientific-based knowledge and, at the same time, enhancing sociocultural change in the perception and role of foreign immigration in rural, mountainous and remote regions of Europe. To clarify the basic and also normative assumptions of this research project, a reflection in a wider public debate through the form of a *Manifesto* as a contribution to EU-level considerations and discourses seems helpful.

Therefore, this book seeks to gain public attention and foster debate at different territorial levels on an innovative and scientifically grounded proposal for the future of a large portion of Europe and its inhabitants. It is based on the declared conviction that this future can and must be built by investing in territorial equity, the enhancement of marginalized areas, the innovative rediscovery of local and regional cultural heritage, the active inclusion of new inhabitants and a radical change of perspective with respect to the current dominance of economic and social agglomeration, "metrophilia" and political centralization.

This book is the very first publication of the new Routledge book series on "Remote Places and Remoteness" proposed and launched by the MATILDE network (editor in chief: Andrea Membretti). Although its authors are scientists, this is not a traditional academic publication. It is intended for both an academic audience and a wider one. In the former case, scholars, researchers and students in different disciplines and interdisciplinary fields (comprising sociology, territorial economics, migration studies, mountain research, human geography, anthropology, demography and political sciences) will be helped by the book's publication to reconsider spatial dynamics, and in particular gain better understanding of the role of remote territories in European development paths, and the importance of migration flows for these processes. In the latter case, the publication will offer concrete proposals for intervention and key elements of reflection to policymakers, journalists, media activists, practitioners, civil servants at different territorial levels and citizens of rural and mountain territories in Europe.

The book therefore intends (i) to provide the means with which to conceptualize the renaissance of remote, mountain and rural places of the continent mainly through policies of attracting and integrating new population and (ii) to contribute to a public debate on concrete local and regional interventions.

The core of the book is constituted by the ten theses that make up the *Manifesto*. They can be summarized as follows, within two main blocks of argumentation:

1) Rural, mountainous and remote regions of Europe are a fundamental but neglected resource for the future of the continent. They need to be radically reconsidered by EU governance and political powers, and a new narrative about them should substitute the current rhetoric of "peripheries," even more so considering the role that these territories can acquire in coping with the COVID-19 pandemic.

The following theses mainly deal with this topic:

Thesis 1. Remoteness needs to be reframed as a resource and place-based value for Europe.
Thesis 2. Rural, mountain and remote regions should be considered as a new core of Europe.
Thesis 3. It is time for a new rural and mountain narrative.
Thesis 8. Rural-urban relationships are fundamental assets in terms of policies aimed at the inclusion of remote places.
Thesis 9. The social and economic development, attractiveness and collective well-being of remote, rural and mountain regions strongly depend on a foundational economy.
Thesis 10. The COVID-19 pandemic can be not only a threat but also an opportunity for remote, rural and mountain regions of Europe, and for their inhabitants.

2) Migration (international and internal) and new peopling movements are among the main drivers of the resilience and future recovery of the *places left behind*. Immigrants (and newcomers in general) must be considered to be, and empowered as, crucial agents of local development together with local inhabitants. Their impact at the socioeconomic level

should be assessed, together with the new challenges posed by migration, in order to advance concrete policies for migrants' valorization and effective inclusion. The following theses mainly deal with this topic:

4 *International migration to rural and mountain areas is an important but neglected phenomenon.*
5 *Migration impact assessment is a powerful tool for local development.*
6 *The inclusion of migrants in rural/mountain territories is a multilevel and multidimensional process.*
7 *International migration has to be considered as one expression among diverse mobilities.*

On the basis of the data produced and their analysis in the MATILDE Project, each thesis aims to discuss a different aspect of the possible, or already-ongoing, renaissance of Europe's remote, mountainous and rural areas. For this renaissance to be possible, environmentally and socially sustainable, as well as be based on a fair redistribution of territorial resources, each thesis highlights the role that new inhabitants can play in these regions, and with a specific focus on foreign immigrants. This role can be active if it is supported by local and supralocal policies. But it also requires negotiation with already-present inhabitants, and it must also integrate needs and options for the inclusion of all types of newcomers, i.e., immigrants, national migrants, returnees, etc., in order to harness their potential and promote innovative alliances and creative synergies.

The presentation of the ten theses is followed by three authoritative comments by internationally known scholars who have long dealt with territorial inequalities, rural and mountain development and international migration issues. These comments should make it possible to extend our views beyond the MATILDE Project with reflection that contextualizes our propositions within the scientific debate and literature of reference.

Finally, the book's concluding chapter takes up the arguments and pathos of the Manifesto in two parts. While the "Symbolic battles for the core of Europe" paragraph examines the people-places nexus in the perspective of the interplay of policies and politics in remote regions, "The dialogical cocreation of living together" restructures the impact of immigration along the axis integration-innovation and governance-citizenship.

This book is primarily aimed at policymakers, local activists, citizens and professionals involved in regeneration projects in marginalized contexts. It is a tool to act and intervene, on the basis of data and scientifically founded analysis, within the framework of a fundamental value orientation: that orientation which sees value in Europe's remote, rural and mountain areas, and in all the citizens who deliberately choose to move to, return to or remain there.

Notes

1 The MATILDE consortium has taken an active part in the open consultation launched by the EC to guide the process of defining and drafting the "Long-term Rural Vision", European Commission (2021) https://ec.europa.eu/info/strategy/priorities-2019-2024/new-push-european-democracy/long-term-vision-rural-areas_en

2 "Dispersal policies" concern the scheme adopted by several European countries to distribute asylum seekers and refugees across the country, sometimes with the aim of counterbalancing negative demographic trends, sometimes in order to reduce pressure on urban centres. This may result in their relocation to relatively disadvantaged areas, where accommodation is cheaper but labour demand is weaker. See Fasani et al. (2018).

References

Alesina, A., Miano, A. and Stantcheva, S. (2018). "Immigration and Redistribution", in *NBER Working Paper 24733*, June 2018. https://www.nber.org/papers?page=1&perPage=50&sortBy=public_date.

Barca, F. (2019). 'Place-Based Policy and Politics', *Renewal: A Journal of Social Democracy*, 27(1), pp. 84–95, London.

Bianchi, M., Caputo, M. L., Lo Cascio, M. and Baglioni, S. (2021). *'A Comparative Analysis of the Migration Phenomenon: A Cross-country Qualitative Analysis of the 10 Country Reports on Migrants' Economic Impact in the MATILDE Regions.* http://doi.org/10.5281/zenodo.5017818. Available at: https://matilde-migration.eu/wp-content/uploads/2021/08/d44-comparative-report-on-tcns-economic-impact-and-entrepreneurship.pdf.

BMI (2020). Territorial Agenda 2030. A future for all places. Informal meeting of Ministers responsible for Spatial Planning and Territorial Development and/or Territorial Cohesion 1 December 2020, Germany. https://ec.europa.eu/regional_policy/sources/docgener/brochure/territorial_agenda_2030_en.pdf.

Caputo, M. L., Bianchi, M., Membretti, A. and Baglioni, S. (eds.). (2021). *10 Country Reports on Economic Impacts* (MATILDE H2020 Project). Available at: https://matilde-migration.eu/wp-content/uploads/2021/07/d43-10-country-reports-on-economic-impacts.pdf.

Charbit, C. (2020). 'From "de jure" to "de facto" Decentralised Public Policies: The Multi-level Governance Approach', *The British Journal of Politics and International Relations*, 22(4), pp. 809–819.

Commission of the European Communities. (2008). *Green Paper on Territorial Cohesion. Turning Territorial Diversity into Strength.* Available at: https://eur-lex.europa.eu/LexUriServ/LexUriServ.do?uri=COM:2008:0616:FIN:EN:PDF.

Copus, A., Kahila, P., Dax, T., Kovács, K., Tagai, G., Weber, R., Grunfelder, J., Meredith, D., Ortega-Reig, M., Piras, S., Löfving, L., Moodie, J., Fritsch, M. and Ferrandis, A. (2021). 'European Shrinking Rural Areas: Key Messages for a Refreshed Long-term European Policy Vision', *TERRA. Revista de desarollo local*, 8, pp. 280–309. http://doi.org/10.7203/terra.8.20366.

Dax, T., Strahl, W., Kirwan, J. and Maye, D. (2016). 'The Leader Programme 2007–2013: Enabling or Disabling Social Innovation and Neo-endogenous

Development? Insights from Austria and Ireland', *European Urban and Regional Studies*, 23(1), pp. 56–68. http://doi.org/10.1177/0969776413490425.
European Commission (2008). Green Paper on Territorial Cohesion, Turning territorial diversity into strength. Communication from the Commission to the Council, the European Parliament, the Committee of the Regions and the European Economic and Social Committee. Document COM(2008) 616 final. Brussels: Commission of the European Communities. https://eur-lex.europa.eu/LexUriServ/LexUriServ.do?uri=COM:2008:0616:FIN:EN:PDF.
European Commission (2021). A long-term vision for the EU's rural areas. Building the future of rural areas together. Brussels. https://ec.europa.eu/info/strategy/priorities-2019-2024/new-push-european-democracy/long-term-vision-rural-areas_en.
ForAlps (n.y.). *Immigration as opportunity for Alpine regions*. Pettinengo, Italy. https://www.foralps.eu.
Galera, G., Giannetto, L., Membretti, A. and Noya, A. (2018). *Integration of Migrants, Refugees and Asylum Seekers in Remote Areas with Declining Populations*, OECD Local Economic and Employment Development (LEED) Working Papers, 2018/0, OECD Publishing, Paris.
Inglehart, R. and Norris, P. (2016). 'Trump, Brexit, and the Rise of Populism: Economic Have-Nots and Cultural Backlash', in *HKS Working Paper No. 26*, Boston, MA.
Kordel, S. and Membretti, A. (eds.). (2020). *Classification of MATILDE Regions. Spatial Specificities and Third Country National Distribution* (MATILDE H2020 Project). Available at: https://matilde-migration.eu/reports/.
Laine, J. (ed.). (2021). *10 Country Reports on Social Impacts Focusing on Qualitative Impacts of TCNs Arrival and Settlement* (MATILDE H2020 Project). Available at: https://matilde-migration.eu/wp-content/uploads/2021/06/D33-10-Reports-on-QualitativeImpacts.pdf.
Lewin, K. (1946). 'Action Research and Minority Problems', *Journal of Social Issues*, 2(4), pp. 34–46. https://doi.org/10.1111/j.1540-4560.1946.tb02295.x.
Magone, J. M. (2003). *Regional Institutions and Governance in the European Union*. Westport, CT: Praeger.
Membretti, A., Kofler, I. and Viazzo, P. P. (eds.). (2017). *Per forza o per scelta: L'immigrazione straniera nelle Alpi e negli Appennini*. Roma: Aracne et al.
Perlik, M., Galera, G., Machold, I. and Membretti, A. (eds.). (2019). *Alpine Refugees: Immigration at the Core of Europe*. Cambridge Scholar Publishing. https://www.mrd-journal.org/.
Perlik, M. and Membretti, A. (2018). 'Migration by Necessity and by Force to Mountain Areas: An Opportunity for Social Innovation', *Mountain Research and Development*, 38(3), pp. 250–264.
Remotti, F. (2011). *Cultura. Dalla complessità all'impoverimento*. Roma: Laterza.
Rodríguez-Pose, A. (2018). 'The Revenge of the Places that Don't Matter (and What to Do about It)', *Cambridge Journal of Regions, Economy and Society*, 11(1), pp. 189–209, Oxford: Oxford University Press.
Shucksmith, M., Brooks, E. and Manipour, A. (2020). 'LEADER and Spatial Justice', *Sociologia Ruralis*, 61(2), pp. 322–343. http://doi.org/10.1111/soru.12334.

2

Ten theses

2.1 Thesis 1

Reframing remote places and remoteness as a collective resource and value for Europe

Andrea Membretti, Thomas Dax and Ingrid Machold

Marginalization as a result of neoliberal globalization and the "revenge of the state"

Neoliberal globalization takes space and opportunities away from people, subordinating them to global networks of capital: it contributes to an accelerated exchange of goods, and thus establishes wide-ranging sociospatial subordination to hegemony power (Harvey, 1985a). Within this economic and political process, physical space tends to be treated as a mere support, an infrastructural platform for the development of productive and service activities that temporarily exploit geographical positions, only to soon abandon them and move to other locations more profitable in terms of labour costs, workers' rights, antipollution laws, taxation, etc. (Jessop, 2000; Amin, 2001). In many respects, concrete and communitarian places are considered by global capitalism to be interchangeable "non-places" (Augé, 1992), where distinctive local features are reduced to those that are functional and convenient for the exploitation of the contexts in which the economic activities are temporary located (Gough, 2014).

These dynamics produce (unstable) territorial hierarchies, competition among territories at global level, shifting alliances between global cities seeking to prevail on the planetary or macroregional economic chessboard. At the same time, these same processes favour the marginalization of large portions of the planet. They affect spaces already exploited and then abandoned by transnational economic actors as they relocate their activities elsewhere. These processes of displacement impoverish local economies and reduce the capacity for self-determination of local communities relying on systems of production that never really take root in their territory but constantly threaten to leave for more convenient locations elsewhere (Gray and Barford, 2018).

At the same time, marginalized areas tend to be both neglected and dispossessed of their prerogatives of self-government also by national political

DOI: 10.4324/9781003260486-4

actors often committed to neoliberal models of development. In fact, while the state has been weakened by the long-lasting economic globalization, on the other hand it now seems reinforced in some respects (e.g., in its control over territorial resources and exercise of power at different scale) by the multitude of crises that have occurred in recent years, like those related to migration, the COVID-19 pandemic and climate change. Therefore, in the emergence of the so-called "neo-post-Westphalian order" in Europe, some see a "revenge of the state" (Krasteva, 2020) that privileges the central power versus local autonomy, despite the old and new claims for more self-determination and control over their own territory advanced by remote localities, even more so amid the ongoing pandemic (OECD, 2021). Pressures on local actors are reinforced by national states and other institutions (like the European Union), powerful economic players, stakeholders, etc. These forces are especially experienced by remote and mountain regions – as clarified by the concept of "mountain pressures" (Klein et al., 2019) for mountain social-ecological systems – thus demonstrating the close dependency of such remote places on extralocal dynamics.

Remote places, between cultural removement and alterity

The word *remoteness* derives from the Latin verb *removeo*, which means **"to remove,"** move something (or someone) away from something else. Remote places are the ones perceived, at least by those not living there, as removed from everyday life, far away (culturally, even more than physically) from the shared social space and the constructed world of meanings that belongs to those societies and groups that define themselves as "the core," the centre of (at least) a large part of the globe. This is what the Europeans have always claimed to be on a global scale, but this cultural and social mechanism does not operate only on an international and wider scale: remoteness is also something perceived by people living in metropolitan areas with respect to some mountain and rural regions of the same state or even the same region represented as remote.

As the anthropologist Edwin Ardener (2012) argues, "the remote is compounded of 'imaginary' as well as 'real' places" (Ibid., p. 521). For Europeans, "remote" areas are conventionally physically "removed," but this obscures the conceptual phenomena associated with "remoteness" (Ibid., p. 522). However, if it is necessary for remoteness to have a topographical location, it "is defined within a topological space whose features are expressed in a cultural vocabulary" (Ibid., p. 523): "it is first of all a conceptual experience" (Ibid., p. 524).

As a consequence, according to Ardener, "remoteness is *a specification, and a perception, from elsewhere*, from an outside standpoint; but from inside the people have their own perceptions – if you like, a counter-specification of the dominant, or defining space, working in the opposite direction" (Ibid. p. 531).

> The lesson of remote areas is that this is a condition not related to periphery, but to the fact that *certain peripheries are by definition not properly linked to the dominant zone. They are perceptions from the dominant zone*, not part of its codified experience. Not all purely geographical peripheries are in this condition, and it is not restricted to peripheries.
>
> Ibid., p. 532

Within marginalization processes driven by neoliberal globalization, remote places therefore seem to constitute a category different from that of "periphery": they represent, in cultural terms, an ***alterity*** posed out of the spatial and conceptual continuum lived and enacted in metropolitan spaces. Remote places can be, and often are, marginalized territories; however, their distinctive feature is neither marginalization nor peripherality, but instead remoteness. This characteristic – attributed mainly through processes of hetero-signification from the "outside world" – has huge potential in terms of innovation, creativity and attractiveness that other territories, perceived as peripheral and marginal, usually do not possess. It is the potential of counter-specification especially apparent when local people are aware of it.

New momentum for remoteness and remote places: climate change and the COVID-19 pandemic

Despite the above-recalled long-lasting processes of territorial marginalization and socioeconomic exclusion which hit mountainous and rural regions particularly hard, it seems that Europe is experiencing unprecedented momentum for remoteness and remote places. Remote areas, and the meaning of the concept of remoteness, are in fact acquiring unprecedented value, on the basis of two global macrophenomena: climate change and the COVID-19 pandemic (see Thesis 10).

Remote areas, and mountains most of all, are often particularly susceptible to the dramatic consequences of climate change: a stronger impact of changes in local temperatures and precipitation; hydrogeological instability and extreme weather events; rains and droughts that hit with particular violence mountains, small islands, inland areas and fragile ecosystems. Among the European macro regions, the Alps are experiencing these phenomena even more dramatically (Schneiderbauer et al., 2021). Outside Europe, in different parts of the globe, these are all factors driving (forced) out-migration from fragile territories, whether temporary or permanent. The consequences are evident in terms of the overcrowding of neighbouring or even distant urban areas, as well as in terms of transregional or international movements of people (often crossing other remote regions located on migration routes), with the further effect of an increasing number of **environmental refugees** (Piguet et al., 2018). As a consequence, many remote places seem to be increasingly affected by striking phenomena of mass **abandonment** and population ageing, with disastrous effects both on local economies and societies and on the possibility of preserving ecosystems often dependent on constant human maintenance and care (Dax et al., 2021; Copus et al., 2021).

Yet it is precisely remote areas – in particular the ones preserving their natural and environmental resources – that in recent years have attracted increasing interest by important sectors of the population in many parts of the world, principally the richest ones of Europe and the Global North. Middle-class and well-educated people from urban areas see them as living spaces away from the "chaos" of the metropolis, with a wide range of ecosystem services provided, and high ecological quality. At the same time, the growing realization that it is possible to work in an innovative manner in these areas has triggered interest in (g)local development (Lardies-Bosque and Membretti, 2022; Barbera et al., 2019).

The ongoing **COVID-19 pandemic** has contributed since 2020 to a rethematization of remoteness, not only in terms of "social distancing" (and the social role of physical distance), but also with regard to perceptions and uses of space – at the physical, symbolic and normative levels. The relationship between "central places" and "marginalized" localities has been attributed new value, albeit not without an ambiguity that reflects socioeconomic disparities among different sectors of the population. For some, remoteness is the attractive counterpart of overcrowded, polluted and often-unsafe metropolises. For others (often local populations remaining in those territories and fragile categories like immigrants), it denotes sociospatial confinement, having to live "far away from" and cut off from services and opportunities, with an even more severe impact in times of lockdown and reduced mobility (Rodriguez-Pose and Hardy, 2015; Perlik, 2011). It therefore happens that the very same place can be perceived as "remote" (with a positive meaning) by newcomers, and as "marginalized" (negatively) by its original inhabitants.

Multiple processes such as migration, climate change, the dissatisfaction of certain social groups with respect to urban life, digitalization, and the return to rurality by the younger generations in some European areas are highlighting how less-anthropized, marginalized and remote territories have an unprecedented potential attractiveness due to their natural resources, the extensive space available and their (relative) distance (at least perceived) from urban areas (Steinecke et al., 2009).

Places matter! Inequalities and opportunities in the regions facing remotization

A reflection on remoteness and its changing cultural attributes needs to start from the **spatial dimension**, recognizing that "place matters" (Dreier et al., 2004; Gieryn, 2000; Massey, 1994), in particular in a globalized world, despite any attempt to represent the world itself as "flat" and the territory as "fluid" and a-dimensional. The spatial dimension and local contexts, in fact, frame societal reproduction and change (Goffman, 1974), shaping continuous sociocultural negotiation among a variety of social structures and groups in territories and involving different groups of inhabitants: established and new, temporary and permanent, nationals and foreigners (Membretti and Viazzo, 2017).

It is the place-based and spatial reflection on remote areas that highlights how different and intertwined factors are accentuating processes of **"remotization"** worldwide in relation to the growing global urbanization and hyperconcentration of socioeconomic activities and households. This happens together with the progressive detachment of many metropolitan areas from their "rural outback" in relation to the dynamics of "disembedded" competition among global cities (Sassen, 2001). According to Membretti (2021), **remotization can be described as the increasing physical and symbolic distance** between and within rural/mountain and urban areas and their populations. This ambivalent process of reciprocal (cultural and physical) **removal and sociospatial rarefaction,** accompanied by a widespread perception of unprecedent remoteness, the widening of everyday living spaces and the stretching/weakening of connections/ties, can lead both to social resentment/isolation and to the opening of new opportunities for local development, innovation and new lifestyles.

Even though new, positive meanings have been attributed to remoteness in recent years, the fact remains that the process of remotization tends often to coincide with marginalization, accompanied by new trends of social exclusion, growing territorial inequalities and widespread resentment expressed – often in terms of populism and xenophobia – by the inhabitants of the "places that do not matter" (Rodriguez-Pose, 2017) against the elites of "central places" (also targeting different scapegoats like immigrants). Although a number of European policies have provided support to rural and mountain regions, perceptions of being "on the margins" of economic and social development are growing stronger within these "remotized" communities (Perlik et al., 2019).

Remotization does not affect only rural and mountainous regions, however. The widespread phenomenon of **shrinking postindustrial regions** and the growing number of cities abandoned by their inhabitants (often due to the collapse of industrial economies) show how urban areas can also become marginalized and increasingly perceived as remote, even cast away, both by their inhabitants and by the outside world (Pallagst et al., 2013; Martinez-Fernandez et al., 2012). In declining cities (e.g., in Eastern EU countries), as in many suburban and rural areas, and especially among the impoverished middle class, resentment towards the central elites is growing, together with unprecedent inequality exacerbated by the COVID-19 pandemic. The perception of being subject to an undesired process of remotization coinciding with sociospatial exclusion is leading to populism, to the "revenge of the places that do not matter" (Rodriguez-Pose, 2017).

At the same time, remotization is a **two-way process** in terms of perception and representation: while the declining regions and the areas physically and symbolically removed perceive themselves as distant from "the centre," it is "the centre" itself (coinciding with the metropolitan and urban areas) that is increasingly perceived as remote (removed) by those living "at the margins." The large metropolitan areas – where political, economic and cultural power is exercised – are beginning to be seen as extremely remote (as regards spatial connections and in terms of imagery) with respect to those regional or **local microcosms** to which their inhabitants are attributing a growing social value as places of refuge, in some cases, or spaces of (resistant/defensive) identity. The **COVID-19** pandemic has even revealed the consequences of a "**compulsion to locality**" – that is, a strong push to remain in specific portions of territory due to lockdown and mobility restrictions but also as a consequence of personal choice (Membretti, 2021), particularly affecting those that are already in most precarious and fragile positions within the society and, thus, experiencing the **local dimension of inequality** more dramatically (Gruber et al., 2022; Kordel and Membretti, 2020; Galera et al., 2018).

However, it is precisely the remote, rural and mountainous areas that throughout Europe are gaining growing interest during the pandemic as **opportunities**, due to better environmental quality and the possibility of living and working in an innovative way on a **"glocal"** scale, using new technologies and **digitalization** (smart working, e-commerce, homeschooling, etc.). Thus, **new population movements** have emerged in terms of in-migration to remote areas by so-called **amenity migrants, new highlanders and neorurals** (Moss, 2006; Membretti and Iancu, 2017; Gretter et al., 2017). While this is happening in some European regions (like the Alps and the Pyrenees), other rural and mountainous territories (e.g., in Eastern Europe and Scandinavia) are still witnessing constant **out-migration** to metropolitan areas (Price et al., 2019). In terms of **migration** processes, remote areas have for several years emerged as new destinations (McAreavey, 2017; see Thesis 7). But a number of emerging types of migration respond to, and are fuelled by, the complexity of contemporary spatial dynamics. In this process, also remote regions increasingly find appropriate "niches" as specific and pulling-in regions.

Thus, it is the ambiguity with respect to the **value and disvalue of remote places** in relation to what they can offer as opportunities for certain social categories in certain places, or with respect to the limits they can impose on other social groups, that merits attention, as does the **concept of remoteness** and the related process of remotization.

Promoting social justice in remote places: a new territorial balance

In recent years, the socioeconomic, cultural, and environmental importance that remote places are gaining has been shown by several studies exploring aspects of inequality, territorial justice, marginalization, and sustainable mountain development (e.g., in the H2020 projects of MATILDE, Welcoming Spaces, RELOCAL, SIMRA). These studies underline the need for further research and innovative actions in order to identify a **different balance** between urban and rural, centres and peripheries, and to furnish a **renewed connection** able to foster social inclusion, widespread well-being, and fair access to the resources of territories for all the various groups of the population interested in their use, within a framework of **sustainable development** and **social justice and equity**.

While considering the old and new socioeconomic disparities increasingly affecting all the regions of Europe, and when seeking innovative governance strategies to tackle these challenges, there seems to be space for **a rediscovery of remote places** in relationship with a **rethematization of remoteness**, as a lens through which to understand some of the most radical changes in everyday life.

COVID-19 and EU recovery plans offer the occasion for **putting remote areas at the centre of the debate on the future of Europe.** This involves new discourses on and approaches to social and territorial justice, territorially balanced and sustainable development. It is an opportunity to **question mainstream views on globalization** that conceal strategies that lead to a dead end by adopting a reversal of gaze: from the "margins of the planet" to what claims to be the centre of the globe.

At the same time, it is an opportunity to **reconceptualize the theme of remoteness**: building on semantic implications and the inherent value of this concept that may be linked to the cultural legacy rooted in local experiences, including inspiring views of indigenous people (Wilder et al., 2016) and avoiding the return of any Eurocentric temptation. These new approaches are fundamental in their search on finding **place-sensitive responses** (Sotarauta,

2020) in order to secure proximity and overcome distance, individual personal development and community life and interlinkages and collaboration, even when living "apart." Importantly, those localized views extend to human/nature relationships, imploring a balance in our complex resource use systems. In comparison to ever-increasing densities of urban and rural places, this altered perspective seems a key requirement.

A balanced view on community life, social contribution and the critical role of local actors is not only a precondition for social and territorial justice but also a foundational element of democratic participation. Remoteness and remote places offer an unexpected and unique opportunity to restore places and spaces to the people, putting territorial equity and translocal solidarity at the very core of "next-generation Europe."

References

Amin, A. (2001). 'Globalisation: Geographical Imaginations', in Smelser, N. and Baltes, P. (eds.), *International Encyclopaedia of the Social and Behavioural Sciences*. Amsterdam: Elsevier Science, pp. 1247–1257.

Ardener, E. (2012). 'Remote Areas. Some Theoretical Considerations', *HAU: Journal of Ethnographic Theory*, 2(1), pp. 519–533. https://doi.org/10.14318/hau2.1.023 (downloaded: 01/12/2021).

Augé, M. (1992). *Non-places: Introduction to an Anthropology of Supermodernity*. Montrouge: Le Seuil.

Barbera, F., Bacchetti, E., Membretti, A., Spirito, A. e Orestano, L. (2019). *Vado a vivere in Montagna. Risposte innovative per sviluppare nuove economie nelle Aree Interne*. Torino: SocialFare.

Copus, A., et al. (2021). 'European Shrinking Rural Areas: Key Messages for a Refreshed Long-term European Policy Vision', *TERRA. Revista de desarollo local*, 8, pp. 280–309. http://doi.org/10.7203/terra.8.20366.

Dax, T., et al. (2021). 'Land Abandonment in Mountain Areas of the EU: An Inevitable Side Effect of Farming Modernization and Neglected Threat to Sustainable Land Use', *Land*, 10(6), p. 591. https://doi.org/10.3390/land10060591.

Dreier, P., Mollenkopf, P. and Swanstrom, T. (2004). *Place Matters: Metropolitics for the Twenty-First Century*. Westbrooke Circle, Lawrence, KS: University Press of Kansas.

Galera, G., et al. (2018). 'Integration of Migrants, Refugees and Asylum Seekers in Remote Areas with Declining Populations', in *OECD Local Economic and Employment Development (LEED) Working Papers, 2018/0*. Paris: OECD Publishing.

Gieryn, T. F. (2000). 'A Space for Place in Sociology', *Annual Review of Sociology*, 26, pp. 463–496.

Goffman E. (1974). *Frame Analysis: An Essay on the Organization of Experience*. Harvard University Press.

Gough, J. (2014). 'The Difference between Local and National Capitalism, and Why Local Capitalisms Differ from One Another: A Marxist Approach', *Capital and Class*, 38(1), pp. 197–210.

Gray, M. and Barford, A. (2018). 'The Depths of the Cuts: The Uneven Geography of Local Government Austerity', *Cambridge Journal of Regions, Economy and Society*, 11(3), pp. 541–563. https://doi.org/10.1093/cjres/rsy019.

Gretter, A., et al. (2017), 'Pathways of Immigration in the Alps and Carpathians: Social Innovation and the Creation of a Welcoming Culture', *MRD – Mountain Research and Development*, 37(4), pp. 396–405.

Gruber, M., et al. (2022). 'The Impact of the COVID-19 Pandemic on Remote and Rural Regions of Europe: Spotlight on Foreign Immigration and Local Development', in *GSSI Discussion Paper series in Regional Science & Economic Geography*, 1/2022, forthcoming.

Harvey, D. (1985a). 'The Geopolitics of Capitalism', in Gregory, D. and Urry, J. (eds.), *Social Relations and Spatial Structures*. London: Macmillan, pp. 128–163.

Harvey, D. (1985b). *The Urbanization of Capital*. Baltimore: Johns Hopkins University Press.

Jessop, B. (2000). 'The Crisis of the National Spatio-temporal Fix and the Tendential Ecological Dominance of Globalizing Capitalism', *International Journal of Urban and Regional Research*, 24, pp. 323–360.

Klein, J. A., et al. (2019). 'Catalyzing Transformations to Sustainability in the World's Mountains', *Earth's Future*, 7, pp. 547–557. http://doi.org/10.1029/2018EF001024.

Kordel, S. and Membretti, A. (eds.). (2020). *Classification of MATILDE Regions. Spatial Specificities ant Third Country National Distribution* (MATILDE H2020 Project). Available at: https://matilde-migration.eu/reports/.

Krasteva, A. (2020). 'If Borders Did Not Exist, Euroscepticism Would Have Invented Them Or, on Post-Communist Re/De/Re/Bordering in Bulgaria', *Geopolitics*, 25(3), pp. 678–705. http://doi.org/10.1080/14650045.2017.1398142.

Lardies-Bosque, R. and Membretti, A. (2022, forthcoming). 'In-migration to European Mountain Regions: A Challenge for Local Resilience and Sustainable Development', in Schneiderbauer, S., Szarzynski, J. and Shroder, J. (eds.), *Safeguarding Mountains – A Global Challenge. Facing Emerging Risks, Adapting to Changing Environments and Building Transformative Resilience in Mountain Regions Worldwide*. Elsevier.

Martinez-Fernandez, C., Audirac, I., Fol, S., and Cunningham-Sabot, E. (2012). Shrinking Cities: Urban Challenges of Globalization. *International Journal of Urban and Regional Research*, 36(2), 213–225.

Massey, D. (1994). *Space, Place and Gender*. Cambridge: Polity.

McAreavey, R. (2017). *New Immigration Destinations, Migrating to Rural and Peripheral Areas*. Abingdon: Routledge.

Membretti, A. (2021). 'Remote Places of Europe and the New Value of Remoteness', *MATILDE: Migration Impact Assessment to Enhance Integration and Local Development in European Rural and Mountain Areas*, September 2021. http://doi.org/10.13140/RG.2.2.15779.78886.

Membretti, A. and Iancu, B. (2017), 'From Peasant Workers to Amenity Migrants. Socialist Heritage and the Future of Mountain Rurality in Romania', *Revue de Geographie Alpine/Journal of Alpine Research*, 105–1.

Membretti, A. and Viazzo, P. P. (2017), 'Negotiating the Mountains. Foreign Immigration and Cultural Change in the Italian Alps', *MARTOR – The Museum of the Romanian Peasant Anthropology Journal*, 22, pp. 93–107.

Moss, L. A., ed. (2006). *The Amenity Migrants: Seeking and Sustaining Mountains and Their Cultures*. Wallingford: CABI.

OECD. (2021). *OECD Regional Outlook 2021: Addressing COVID-19 and Moving to Net Zero Greenhouse Gas Emissions*. Paris: OECD Publishing.

Pallagst, K., Wiechmann, T., and Martinez-Fernandez, C. (eds.). (2013). *Shrinking Cities: International Perspectives and Policy Implications*. Routledge: London.

Perlik, M. (2011). 'Alpine Gentrification: The Mountain Village as a Metropolitan Neighbourhood', *Revue de Geographie Alpine/Journal of Alpine Research*, 99(1), http://doi.org/10.4000/rga.1370.

Perlik, M., Galera, G., Machold, I. and Membretti, A., (eds.). (2019). *Alpine Refugees. Immigration at the Core of Europe*. Cambridge: Cambridge Scholars Publishing.

Piguet, E., Kaenzig, R. and Guélat, J. (2018). 'The Uneven Geography of Research on "Environmental Migration"', *Popul Environ*, 39, pp. 357–383. https://doi.org/10.1007/s11111-018-0296-4.

Price, M. F., Arnesen, T., Gloersen, E. and Metzger, M. J. (2019). Mapping Mountain Areas: Learning from Global, European and Norwegian Perspectives. *Journal of Mountain Science*. 16(1): 1–15.

Rodríguez-Pose, A. (2017). 'The Revenge of the Places that Don't Matter (and What to Do about It)', *Cambridge Journal of Regions, Economy and Society*, 11(1), pp. 189–209, Oxford: Oxford University Press.

Rodriguez-Pose, A. and Hardy, D. (2015). 'Addressing Poverty and Inequality in the Rural Economy from a Global Perspective', *Applied Geography*, 61, pp. 11–23. http://doi.org/10.1016/j.apgeog.2015.02.005.

Sassen, S. (2001). *The Global City: New York, London, Tokyo*. Princeton: Princeton University Press.

Schneiderbauer, S., Szarzynski, J. and Shroder, J. (eds.). (2021). *Safeguarding Mountains – A Global Challenge. Facing Emerging Risks, Adapting to Changing Environments and Building Transformative Resilience in Mountain Regions Worldwide*. London: Elsevier.

Sotarauta, M. (2020). 'Place-based Policy, Place Sensitivity and Place Leadership', *Working Paper 46/2020*. Tampere University.

Steinecke, E., Čede, P. and Flie, U. (2009). 'Development Patterns of Rural Depopulation Areas. Demographic Impacts of Amenity Migration on Italian Peripheral Regions', *Mitteilungen der Osterreichischen Geographischen Gesellschaft*, pp. 195–214.

Wilder, B. T., et al. (2016). 'The Importance of Indigenous Knowledge in Curbing the Loss of Language and Biodiversity', *Bioscience*, 66(6), pp. 499–509. http://doi.org/10.1093/biosci/biw026.

2.2 Thesis 2

Rural, mountain and remote regions should be treated as the core of Europe, and the role of migration need to be considered for recovery and development

Susanne Stenbacka and Ulf Hansson

Introduction

Rural and mountain regions can be considered (in many respects) as the core of Europe – that is, in the sense that these areas are vital for the functioning of the European continent. A bird's-eye view of Europe and its diversity of countries makes the region's geographical structure visible. It shows its vast extent and areas where upland as well as lowland landscapes meet mountain areas, but also large cities and their surroundings. The network that connects all these places is both visible and invisible. It is physically visible in the form of roads, railways and airports, and invisible in communication links such as individuals' networks, thoughts and intentions. The transformation of rural areas from productivist to multifunctional places has been both acknowledged and promoted since the late 1990s. In general, it means emphasizing rural diversity, from a single focus on agriculture to acknowledging the rural as "post-productivist" (Wilson, 2001). Examples of post-productivist indicators are in line with what Wilson and Rigg (2003, p. 681) refer to as "policy change; organic farming; counter-urbanization; the inclusion of environmental NGOs at the core of policy-making; the consumption of the countryside; and on-farm diversification activities." Markey et al. (2008, p. 411) emphasize that rural areas, while struggling with certain challenges, "contain many assets that are highly valued within the contemporary global economy, such as: access to resources, natural amenities and high quality of life, and inexpensive land." Furthermore, climate change means that interest in the countryside is increasing even more, and it accentuates the importance of the countryside as a producer of raw materials such as wood and energy, but also food and recreational environments. Rural areas are, for example, crucial for leading the world to a bio-based and fossil-free economy (SOU, 2017, p. 1).[1] In this chapter, we discuss how

DOI: 10.4324/9781003260486-5

the function of rural areas should be understood in a context that comprises not only how these areas provide resources but also their access to resources and how international migration relates to this relationship. In this endeavour, we proceed by first elaborating on the concept of rural stress, its content and how it might relate to international migration.

Stress in rural areas may stem from the imminent threat of the downsizing of services and cuts in welfare provision. Another source of stress is the increased expectation on rural areas functioning as a supplier of energy. The physical environment transforms through the presence of, for example, windmills, and tensions or disagreements might arise locally when different standpoints meet with regard to the impact on the landscape. Stress will eventually make it difficult to provide these assets and functions, whether they be energy production, recreation or people's health. However, in specific and demanding situations, stress can generate extra strength and energy. In a rural village, this can be expressed in actions intended to preserve or transform the place. Such aspects also intersect, because mobilization to save a school, for example, or demonstrations against agricultural reforms may be means to change the future, as well as to preserve something that already exists. Similarly, a mobilization to prevent the deportation of immigrants who are denied asylum and who have settled in a village could be seen as affirmation of something new, and where maintaining a certain population number is necessary to preserve a certain structure. Thus, situations causing stress may, in certain circumstances, constitute a point of departure for dynamic processes.

Situational stressors – structures and room for manoeuvre

There is a broad range of stressors that may arise and be experienced in a rural context. For this reason, it might not be appropriate to speak about rural stress but rather to think about stress in a rural context, "implying a particular rural manifestation of more general stress" (Lobley et al., 2004, p. ii). According to the *Rural Stress Review* (Lobley et al., 2004), the same stressors may occur in diverse contexts, but people living in remote rural communities may experience stress differently because of their stoical outlook and cultural norms, being content with a lower level of support involving limited help-seeking when, for example, experiencing deteriorating health or finances. Gender and occupation are factors that affect subjective experiences of stress. It has also been pointed out that deep-seated structural factors (such as socioeconomic class) affect stress experiences (Lobley et al., 2004, pp. ii–iii). Out-migration and a decline in local support resources, caused by an economic crisis, may generate feelings of social isolation and hopelessness among the community (Lobley et al., 2004, p. 89). In the book

Stress in the European Union, the authors argue that new governance structures and a shifting balance of power cause serious stress at national and international levels (Cramme and Hobolt, 2014). Input factors may be a reduced scope for action and the perception that decisions are made further and further away – as also discussed in terms of remotization in Thesis 1.

Issues that unite many of the individuals who live their lives in rural areas are ones that concern access to public and private services. Political decisions and the allocation of resources affect the distribution of public services such as healthcare and schools, and the prerequisites for the establishment of private services. Failure to make financial investments and thereby – indirectly – convey images of places and regions that are not worth investing in affect both the current life and the belief in the future. Woods (2016) argues that the inability among national governments to deal constructively with the problems associated with rural areas has led to disillusionment in many rural places (Woods, 2016, p. 627). Disillusionment and resignation to the fact that a place's resources are increasingly limited have also been found to be followed by a decline in welfare services related to safety and security (Stenbacka, 2022). In addition, experiences of economic and political dependency on decision-makers in the metropolitan centres may contribute to increasing the symbolic distance to the city and the political elite (Cramme and Hobolt, 2014; Stenbacka, 2020).

The relation between stress and social cohesion can be understood by focusing upon individuals and/or structures. Before the COVID-19 pandemic, "stress" was primarily related to gainful employment. However, in the EU, the pandemic's effects on mental health have received attention. The report "Health at a Glance: Europe 2020" discusses evidence that there are higher rates of stress, anxiety and depression, in particular among specific groups, and that while the stressors may be common and shared, the experiences differ. Some observers label this a "second" or "silent" pandemic (Scholz, 2021). Groups that are mentioned are, for example, elderly persons, children and adolescents and healthcare workers. Stress can thus be identified and discussed according to different human contexts; it may affect individuals, families or households and communities. Among researchers using the framework of resilience, it is common to differentiate between rapid catastrophic trauma and slow-paced traumatic stress. The latter, also labelled "slow distress" (Yamamoto, 2011), refers to long-lasting conditions. Examples of slow distress are prolonged industrial restructuring, ageing population and infrastructure deterioration – disturbances that are cumulative, insidious and endogenous (Foster, 2006).

Rural areas have experienced longer periods of financialization of the economy (Hansen, 2021), such as the privatization of services and infrastructure, in parallel with the downscaling of such services. These processes may also be defined as a substantial challenge or threat for well-functioning rural areas. In MATILDE, we have identified disappointment over cuts in public services and public noninvestment (Mathisen and Stenbacka, 2021). The demand for society to mobilize when receiving a high inflow of immigrants was initially accompanied by both state support and civil society mobilization. The reasons for this mobilization can be related to humanism and caring for others, a desire for a growing population, and local dynamics, such as an increase of schoolchildren. The stress was thus counteracted partly by government financial injections (grants) and partly by the feeling that the engagement of working with immigrants had both an immediate and a more long-term significance. In some cases, public service expanded during the period of high refugee reception, but when the establishment period was over, resources decreased. On the local level, networks grew strong both among and between public institutions and civil society organizations. However, doubts were expressed, also on a local level, with regard to the national policies. The territorialization of what is experienced as social injustice in terms of access to material and symbolic resources can contribute to increased local cohesion in parallel with weak regional or national cohesion (Stenbacka, 2022).

Stress in rural areas thus impacts on both societies and individuals, and there is a risk that the countryside may lose its ability to thrive and develop when the prerequisites change or when a lack of state support and recognition is experienced. The significant value potential found in the physical environment, a resilient and adaptable population and a social small-scale context that may enhance integration needs a basic structure to be meaningful. Such a basic structure entails, in line with the foundational economy approach (also discussed in Thesis 9), the material domain and the providential domain. The material domain consists of, for example, water, electricity, transportation, communication and food; and the providential domain refers to welfare services, including education, health and social care and public administration – services available to all citizens (Nygaard and Hansen, 2020). If mountain and rural areas are to function as the alternative core of Europe, such fundamental conditions should be fulfilled.

The role of migrants in counteracting stress in rural areas

As a consequence of events in 2015 and 2016, countries all across Europe experienced large-scale immigration. In this chapter we consider the case of Sweden, a country that received a large number of immigrants (more than most other European countries). Immigration has led to population increases in remote and rural areas. Furthermore, it has also brought with it an increase in government funding for housing, education and integration.

In terms of demographic changes, one particular aspect highlighted in previous studies is that immigration may rejuvenate ageing rural communities suffering from youth out-migration and population ageing, because immigrants often have younger age structures than native-born populations (Hedberg and Haandrikman, 2014). There may also be the case that immigration helps to balance female out-migration, thereby counteract unbalanced sex ratios, and reduces fertility rates in rural areas, because younger immigrant women are drawn to rural areas as new labour or marriage migrants (Hedberg and Haandrikman, 2014). As highlighted elsewhere – see Lardiés-Bosque and Del Olmo-Vicén (2021) regarding the situation in Aragon (Spain) and Spenger et al. (2021) regarding Austria – Sweden is not unique in this respect. International migration has had a positive influence on the demographic balance in Sweden, though not necessarily in terms of gender balance. As highlighted in the case of the county of Dalarna, for example, between the years 2008 and 2018, the number of foreign-born residents increased from 21,893 to 37,163, an increase in the share of foreign-born citizens from 7.9% in 2008 to 12.9% in 2018.

Municipalities may see receiving refugees and labour migrants as an important strategy for creating a resilient economic and social environment. Thus, rural places cannot be viewed as passive recipients or defenders of the local but as agents of globalization (Stenbacka, 2013). The revitalization and rejuvenation of sparsely populated areas through immigration is a narrative discussed by several researchers. Westholm (2016), for example, writes that the arrival of third-country nationals/migrants can be seen as an injection for society at large and whereby, for example, dormant clubs and societies have been able to find new members. People open up to help the new residents. Shops obtain new customers who ask for other goods. This has given hope for the future not just to the refugees arriving in these small communities but also to the small rural municipalities themselves. Similarly, Hedberg (2016) refers to the contribution by migrants to a "dynamic countryside" where local residents appreciate – in the case of Thai berry pickers – "an international presence" as well as the "bringing of life onto the streets" and adding "a feeling of globalization" to the countryside. She also adds a cautionary point: when immigrants arrive, be they asylum seekers or labour migrants, society needs to focus on the incoming individuals preferably through an established reception process, but also through validation of schooling and previous work.

While it is possible within the just-described context to refer to an injection and revitalization of the countryside in terms of a changed demographic, as well as to the issues highlighted above, it is also necessary to refer to the challenges related to immigration and integration. Vallström (2020), for example, refers to immigrants who encounter stronger and deeper socioeconomic and geographical segregation. They have been allocated apartments left empty as a result of population decline and in areas already experiencing disadvantage with increased pressure on welfare and education systems. These also tend to be areas already disadvantaged and in which municipalities tend to offer reduced services and fewer resources. This was echoed in a research carried out under the remit of the MATILDE Project, which recorded a sensation among the interviewees that rural areas might not always be well equipped to fulfil the needs of the population. In one particular municipality, young people were seen as existing "outside of society" because they lacked education and opportunities for work (Mathisen and Stenbacka, 2021). Interviewees at both local and national levels reflected on this issue of access to resources and wondered whether it might induce people to draw simple conclusions where migrants were used as scapegoats for unwanted developments.

The importance of individuals and structures for recovering from rural stress

Population decreasing and ageing are among the greatest contemporary challenges facing rural, mountainous European areas today. Insufficient infrastructure, e.g., healthcare facilities, educational services and communication means, including public transport and broadband, leads to decreased well-being and dissatisfaction among the inhabitants concerned. So far, the material conditions, in addition, social constructions of regions or places, contribute to creating diverse kinds of distance, in social and cultural terms (Eriksson, 2010). In-migration of people from the world outside to areas sometimes facing population loss and an infrastructure that risks disassembling constitute an injection. The effect of the injection can be hindered by rural stress, but it is possible to recover from stress. Within the rural context, this would involve sufficient infrastructure and institutions that form the basis of people's ability to develop and become part of a society.

The perception of rural areas as crucial for leading the world into an environmentally sound economy, as an environment that contains the assets needed to provide for increased sustainability, is gaining ground. Rural areas provide raw material for food production, construction and energy, as well as areas for recovery and well-being. Such activities and businesses are place-bounded and run by humans. People are thus central to the future of rural Europe, and research as part of the MATILDE Project has highlighted what a demographic "refill" means for the ability to provide certain services such as schools and shops and to fill vacancies necessary for performing work in the healthcare sector. While it is the people that constitute the basis for providing public and private services and engagement in civil society, governmental structures that allow for connections and opportunities are crucial. Material and symbolic functions of the rural are still to be managed in a sustainable manner.

Note

1 However, it is disputed whether such uses really lead to increased sustainability, because, in parallel, they can mean reduced areas for food production and energy-intensive consumption of rural environments (McCarthy, 2008).

References

Cramme, O. and Hobolt, S. B. (eds.). (2014). *Democratic Politics in a European Union under Stress*. Oxford: Oxford University Press.

Eriksson, M. (2010). '"People in Stockholm Are Smarter than Countryside Folks": Reproducing Urban and Rural Imaginaries in Film and Life', *Journal of Rural Studies*, 26(2), pp. 95–104.

Foster, K. (2006). *A Case Study Approach to Understanding Regional Resilience*. Berkeley, CA: Institute of Urban and Regional Development, UC Berkeley. Available at: http://igs.berkeley.edu/brr/workingpapers/2007-08-foster-regional_resilience.pdf (Accessed 18 May 2021).

Hansen, T. (2021). 'The Foundational Economy and Regional Development', *Regional Studies*, pp. 1–10. https://doi.org/10.1080/00343404.2021.1939860.

Hedberg, C. (2016). 'Migration skapar en dynamisk landsbygd', *Kungliga Skogs- och lantbruksakademiens tidskrift*, No. 5, pp. 12–15.

Hedberg, C. and Haandrikman, K. (2014). 'Repopulation of the Swedish Countryside: Globalisation by International Migration', *Journal of Rural Studies*, 34, pp. 128–138.

Lardiés-Bosque, R. and Del Olmo-Vicén, N. (2021). 'Spain', in Laine, J. (ed.), *10 Country Reports on Qualitative Impacts of TCNs* (MATILDE Deliverable 3.3), April 2021, pp. 126–144. http://doi.org/10.5281/zenodo.4726645.

Lobley, M., Johnson, G., Reed, M., Winter, M. and Little, J. (2004). *Rural Stress Review*. Centre for Rural Research. Exeter: University of Exeter.

Markey, S., Halseth, G. and Manson, D. (2008). 'Challenging the Inevitability of Rural Decline: Advancing the Policy of Place in Northern British Columbia', *Journal of Rural Studies*, 24(4), pp 409–421.

Mathisen, T. and Stenbacka, S. (2021). 'Sweden', in Laine, J. (ed.), *10 Country Reports on Qualitative Impacts of TCNs* (MATILDE Deliverable 3.3), pp. 145–162.

McCarthy, J. (2008). 'Rural Geography: Globalizing the Countryside', *Progress in Human Geography*, 32(1), pp. 129–137.

Nygaard, B. and Hansen, T. (2020). Local Development Through the Foundational Economy? Priority-setting in Danish Municipalities', *Local Economy*, 35(8), pp. 768–786.

Scholz, N. (2021). *Mental Health and the Pandemic. Briefing. European Parliament. EPRS, European Parliamentary Research Service*. Members' Research Service PE 696.164 – July 2021.

Spenger, et al. (2021). 'Austria Country Report', in Kordel, S. and Membretti, A. (2020). *Report on Conceptual Frameworks on Migration Processes and Local Development in Rural and Mountain Areas* (Deliverable 2.4 of MATILDE project), pp. 65–132.

Statens Offentliga Utredningar, SOU 2017:1. (2017). [Swedish Government Inquiry] *För Sveriges landsbygder – en sammanhållen politik för arbete, hållbar tillväxt och välfärd.* Stockholm: Elanders AB.

Stenbacka, S. (2013). 'International Migration and Resilience – Rural Introductory Spaces and Refugee Immigration as a Resource', in Tamasy, C. and Revilla Diez, J. (eds.), *Regional Resilience, Economy and Society – Globalising Rural Places.* Oxon: Routledge, pp. 75–94.

Stenbacka, S. (2020). 'Polisfrånvaro och periferialisering', in Stenbacka, S. and Heldt Cassel, S. (eds.), *Periferi som process. Ymer 2020, Årgång 140.* Stockholm: SSAG, Svenska Sällskapet för Antropologi och Geografi, pp. 109–130.

Stenbacka, S. (2022). 'Rural Policing – Spaces of Coherence and Fragmentation', in Bowden, M. and Harkness, A. (eds.), *Rural Transformations and Rural Crime: International Critical Perspectives in Rural Criminology*. Bristol: Bristol University Press, Research in Rural Crime Series, forthcoming.

Vallström, M. (2020). 'Migration, segregation och pauperisering i det svenska samhällets periferi. Kartläggning av tre landsbygdskommuner', *Kulturella Perspektiv. svensk etnologisk tidskrift*, 29(1), pp. 52–61.

Westholm, E. (2016). 'Framtidstro och framtidsoro i flyktingfrågan', *Kungliga. Skogs- och lantbruksakademiens tidskrift*, 5, pp. 8–11.

Wilson, G.A. (2001). From productivism to post-productivism ... and back again? Exploring the (un)changed natural and mental landscapes of European agriculture. *Transactions of the Institute of British Geographers*, 26(1), pp. 77–102. Doi: 10.1111/1475-5661.00007.

Wilson, G. and Rigg, J. (2003). 'Post-Productivist Agricultural Regimes and the South: Discordant Concepts?', *Progress in Human Geography*, 27(6), pp. 681–707.

Woods, M. (2016). 'Confronting Globalisation? Rural Protest, Resistance and Social Movements', in Schucksmith, M. and Brown, D. L. (eds.), *Routldge International Handbook of Rural Studies*. New York: Routledge, pp. 626–637.

Yamamoto, D. (2011). 'Regional Resilience: Prospects for Regional Development Research', *Geography Compass*, 5(10), pp. 723–736.

2.3 Thesis 3

It is time for a new rural and mountain narrative

Thomas Dax, Cristina Dalla Torre and Ingrid Machold

Narratives: structuring a specific vision of reality

There is a perception of persistent "gaps" between the socioeconomic performance of rural and mountain areas and that of urban or agglomeration contexts. This view increasingly upsets local activists raising questions on how "effective" policy adaptation for these areas could be brought about to fill this gap. A discussion in the aftermath of the European election results revealed the increasing "discontent" of local voters in remote rural places – a feeling among local inhabitants that they live in "places left behind" and which do not matter (Rodríguez-Pose, 2018). Although this refers to declining places with a past manufacturing history, it would also apply to remote rural places that are generally losing vitality due to many factors. To some extent, these electoral reactions were seen as responses driven by emotions, perceptions and feelings of "exclusion" among rural people. It appeared as if a convincing negative narrative was built around personal and community experiences in those places.

Narrative is one of the key modes of knowing for human beings, who can be recognized as *homo narrans* (Fisher, 1985), in contrast to former assumptions of seemingly rational decisions driven by features of the *homo economicus*. Humans learn about, make sense of and act in the world through "stories." It is through narrative structures that human beings think, perceive, imagine, make moral choices and create "meaning." Wittmayer et al. (2019) distinguish between narratives that are constructed "top down" and those that are formed in more or less participatory manner by individuals and communities. But independently from the process of construction, narratives have a performative dimension because they carry value, reinterpret the past, and guide current actions in anticipation of a different future. Hence, they can also be seen as part of the wider process of social construction of reality.

However, this is not a homogenous process. Depending on societal discourse patterns, a plethora of divergent narratives compete with each other,

DOI: 10.4324/9781003260486-6

evolving over time, places, social structures and different levels of analysis. "These dominant cultural narratives, even if they are very negative, remain so powerful that despite their own desire to escape from them it is difficult to find alternative personal or community stories to replace them" (Rappaport, 1995, p. 803).

Mainstream narratives as obstacles to socially needed transformation

Currently, mainstream narratives adopt simplistic views on the socioeconomic potential of rural regions, or they focus on romantic perceptions of an idyllic rural life. The problem with these widely shared defensive perspectives is their presentation of stereotypes as the inevitable fates of rural areas (Dax, 2014), attributing almost no decision-making power to rural regions. They instead depict them as devoid of attractive social and economic opportunities and may thus induce populist and reductionist visions of the local space. Indeed, remote rural areas, like many mountain contexts, are experienced as marginalizing. This happens when "urban" assessment standards conceive rural contexts largely as peripheral places. At the same time, there arises closer attention paid to remote regions by specific categories of people, such as "new highlanders" and amenity migrants (see Theses 1 and 5).

The main problem with the hindering features of the prevailing narratives is not just their backward orientation; it is particularly the impetus that they give to "downward-spiralling" processes. Inherently, this creates separation and the hierarchical dependence of mountain and rural population on urban paradigms, depleting their potential and capacity to change. It maintains and reinforces the existing power relations in societies and does not touch upon spatial dependence, inequality, and the pressure on human-nature relationships.

Hence, the most common features of present-day positions on rural/mountain areas are:

- *Seeing rural amenities as strengths*. Rural areas in general, and mountain places in particular, have been assessed as places of "harmony," with a multitude of amenities and assets that shape them as attractive locations. This flattering and overstated attribution includes the notion of extended resilience and propounds the return to rural areas as a creative way out of urban predominance. Presenting rural areas as rich in amenities may enhance rural resilience particularly in times of crises, and it contributes to attracting in-migrants of various origins. On the other hand, the renewed interest in rural and mountain places amid the ongoing COVID-19 pandemic may also reflect a pragmatic assessment with regard to living standards and freedom of movement. Rural areas have been perceived as more resistant and persevering against deprivation and poverty compared to urban areas, where the concentration of poverty is more visible (Woods, 2005, pp. 268–269). However, this overstated positive view hides the risk of not considering instances of underemployment, social deprivation and proletarization in rural and mountain areas.

- *Polarization between peripheries and core.* This aspect appears as the other side of the coin, i.e., the strong dependencies of peripheries on thriving centres, which are perceived as "engines of growth." The widespread approval of this concept is supported by decisive features of peripheralization discourses (see also Thesis 1, process of remotization): those discourses are exacerbated by the "badmouthing" or disparaging remote regions, which means that negative views are repeated more often than inspiring initiatives. This attitude is expressed through the use of "fuzzy language" driven by the complex situation of uneven processes and the neglect of social inequalities within peripheries (Pugh and Dubois, 2021). Such adverse discourses and challenges for remote places are fuelling stereotypes and downward trends instead of combining forces to alter perceptions and focus on alternatives (Dax, 2014).
- *Rural areas as loci for social innovation.* Narratives promoting a normative approach to (social) innovation are functional for the neoliberal growth paradigm. The mainstream discourse places a heavy burden on innovation by local actors and increasingly calls for social innovation in support of market mechanisms. Responsibility for change is transferred to local actors so that spatial reform processes often turn to (implicitly) legitimate neoliberal public policies. In contrast, social innovation as invoked by political and social movements at a local or regional level could go beyond mere adaptation modes and aim at fundamental changes in human relationships to tackle social and territorial inequalities (Moulaert and MacCallum, 2019), proposing new ways of doing and organizing, of framing and knowing.
- *Celebrating individual heroes for the community.* In search of effective rural policies and taking account of the tremendous diversity of rural regions, good practice in local action and policy implementation is often presented as a key element in learning models (Galera and Baglioni, 2021). It tends to epitomize individual actions as core responses at the expense of community engagement and enhanced participation. Even though participation has been an important feature in rhetoric for local and regional policy and particularly in LEADER approaches, the concern to achieve measurable (short-term) outcomes of rural policies has prevailed. Swift project elaboration and a focus on quantitative evaluation have been the prime arguments in building

a narrative of individual agency, while the local capacity to cooperate and achieve success through place-based action has been given less consideration.

- *Neglected places as socially homogenous, ageing regions with no future.* Despite the strong attractiveness of some (accessible) rural and mountain regions, peripheral places are treated as residual areas and tend to be overlooked or just "compensated" in public policies. The aforesaid features concur with a view of peripheral regions as places of out-migration, leaving an ageing society with little capacity to innovate and integrate into modern economy and cultural life (Copus et al., 2020). Even if young people are perceived as a diminishing social group that can see a personal future only outside mountain areas, regional analysis, in general, fails to address specificity, divergence and stimuli for new life approaches and attractive initiatives for young people and incomers. Narratives adopt homogenous perceptions of work and life contexts, neglecting seeds for change (Machold and Dax, 2017).

New trends and challenges for rural and mountain regions

Growing challenges for mountain and rural areas – ranging from social crises to those concerning climate change or resource depletion – urge a large-scale transformation of policies and action, at all levels. Some trends of change started a long time ago; others, such as "new immigration destinations," have gained momentum only recently (see Thesis 7). The assumption here is that a simple juxtaposition of what is considered old and new oversimplifies the complex interrelations among economic, sociodemographic, cultural and environmental factors. These factors do not occur only within rural and mountain areas; they are also linked to other areas with which rural and mountains are connected (e.g., urban, periurban, global).

Besides climate change effects, digitalization and technological innovations, rural and mountain regions have been deeply affected by global socioeconomic changes, such as economic liberalization and tertiarization and related policies, demographic growth, increase in goods and people's mobility.

A progressive penetration of global economic interests at the local level determines every aspect of rural life, and the interdependency of rural/mountain and urban areas is apparent in many aspects of economic development. Global penetration on a local scale engenders restructured market relationships and accelerates both the abandonment of traditional economic activities and the selective intensification of resource exploitation in response to market signals. It thus increases the disparities among and within less-attractive and profitable territories (Jodha, 2000). On the other hand, from a resource dependency point of view, the megatrend of urbanization means that the needs of cities for resources and ecosystem services can rarely be satisfied locally; rather, they depend on extensive and often-distant biosphere support areas, including mountain areas especially.

As a consequence of the megatrend of urbanization, remote areas are faced with depopulation, ageing of the population and a "brain drain" in less attractive places (often high-altitude, nontouristic villages) as highly skilled people "escape" to cities and urban lifestyles. However, in recent decades, many mountain regions have also experienced increases in goods and people's mobility in the form of both in- and out-migration (e.g., seasonal migration, permanent rural exodus, amenity immigration and the incoming of "new highlanders" [Bender and Kanitscheider, 2012]), because sociocultural and technological changes facilitate greater geographical and social mobility.

The above-described demographic, socioeconomic and mobility trends also impact on gentrification processes associated with sociocultural, land-use and housing-related conflicts in rural communities (see Thesis 7). They affect rural property markets, generating inflation in real estate values, including the value of agricultural land. Simultaneously, they open up spaces for negotiation on rights to resource use between "new" and "old" inhabitants, between global needs and local ones. The new ideas expressed in these change processes result in cultural change and a rearrangement of social and human resources. At the same time, they may provoke conflicts over land use and access to resources, power imbalances, and the distribution of access to resources.

As a consequence, rural and mountain regions are confronted with a pressing need for socio-ecological transformation, a need exacerbated by demographic change, economic structural adjustment, biodiversity loss and a growing number of long-term social-ecological challenges (Klein et al., 2019). Those who address these transition demands as separate processes fail to understand their interlinkages and dependence on solidly anchored power relations. This applies at all policy levels.

Designing provident and promising narratives

The commitment to revising and adjusting policies in such a way that they actually enhance the development of rural and mountain regions is hence increasingly discussed in policy circles and by experts as well. This is most evident in the current high-level concern for "Functional Rural Regions" in EC and OECD expert discourses. Grasping opportunities to revaluate rural regions has gained especial importance in discussions during COP26 in Glasgow in November 2021 and in the Long-Term Vision for Rural Areas approved in June 2021 by the EC Commission.

As many studies have underlined, policy reform is not shaped solely by good ideas and conceptual frames devoid of public engagement. It also has to interact with societal needs and discourses. Elaborating "alternative narratives" is therefore an initial step in policy processes, but it is easily neglected because of short-term aspirations for, and exaggerated expectations of, "immediate" results. The selection of appropriate narratives will therefore have to be based on the following aspects:

- Awareness, engagement and participation of local people to enhance community cooperation and integration, and thereby construct and shape place-sensitive views, strategies and visions (narratives of change).
- Acknowledge the complexity of socioecological systems and incite action to enable the regeneration of space.
- Enhance regenerative relations between rural/mountain and urban areas that focus on the diversity of people and places, and interdependence of spaces, as well as the consequences for the attractiveness of places.
- Consider place specificity as central, appreciating the uneven sets of opportunities for diverse groups of people (incomers, local people, out-migrants).
- Reflect on place-specific challenges and reconceptualize rural/mountain spaces as places of destination and not just of departure.
- Renegotiate space by discussing potential conflicts due to change in mobility, economy, culture, climate and transition, but also include perceptions of opportunity (e.g., narratives on systemic design, circularity, decolonialization and the role of commons).

This process of searching for and constructing alternative narratives requires that the *status quo* be questioned and reframed by challenging and confronting dominant norms, values and beliefs. These "alternative narratives" stress the following evolving aspects: the crucial role of interdependencies; new views on functional perspectives and interaction for rural places (Membretti, 2021); participation by large groups of local inhabitants, including recently arrived migrants and newcomers and care for nonhuman elements and natural resources, which seems decisive.

The crucial aspect is to generate authentic transdisciplinary discourse and action that extends beyond knowledge and policy exchange in silos. For this reason, the MATILDE Project aims at producing shared narratives by means of a multistakeholder approach and participatory action research in local case study regions that involves broad discussion by local and regional actors of different backgrounds on local and regional opportunities and challenges. Only then will it be possible to create new narratives that reflect the place specificity of mountains and the needs of local actors.

Deliberately shaping new narratives

Policies and action for rural development and enhancement of mountain regions are deadlocked when they remain in predominant polarizing development frameworks. The shift towards a better integration of emerging needs has proved to be the decisive change to enhance the action and performance of these undervalued areas. Applying place-specific narratives, it is suggested to reflect the changes in mindsets and advance a promising framework for strategies and action in these contexts. The renewed perspective would also encompass settings of greater attractiveness and places of community building suitable for new immigrants. Rural and mountain places would be presented as:

- Places that conceive attractive living modes, considering in a balanced way challenges and opportunities linked to location. This may involve a larger number of in-migrants and strategies to involve them without great delay in community life.
- Places that enable the empowerment and participation of all inhabitants disregarding or modifying existing power relations. Defensive perspectives of rural areas should be transformed into place-based narratives of "self-efficacy."
- Places not seen as stand-alone locations but rather considered in their interdependence with urban contexts. Rural/urban exchange would be captured as a fluid interchange of diverse functions.

It is important that such narratives use a positive language carefully reconsidering the emerging options and presenting visions for a "good life" in rural and mountain places. They would focus on their own deliberative, internally discussed and developed and iteratively constructed narratives of change. These would be built on social innovation and an understanding of transformation oriented to empowerment, participation and community development which enhances local strengths and is open to new ideas and exchanges with other places.

References

Bender, O. and Kanitscheider, S. (2012). 'New Immigration Into the European Alps: Emerging Research Issues', *Mountain Research and Development*, 32(2), pp. 235–241. https://doi.org/10.1659/MRD-JOURNAL-D-12-00030.1.

Copus, A., et al. (2020). *Final Report. European Shrinking Rural Areas: Challenges, Actions and Perspectives for Territorial Governance* (ESPON 2020 project ESCAPE. Version 21/12/2020). Luxembourg: ESPON EGTC. Available at: www.espon.eu/sites/default/files/attachments/ESPON%20ESCAPE%20Main%20Final%20Report.pdf (Accessed 10 December 2021).

Dax, T. (2014). 'A New Rationale for Rural Cohesion Policy: Overcoming Spatial Stereotypes by Addressing Inter-relations and Opportunities', in OECD (ed.), *Innovation and Modernising the Rural Economy*. Paris: OECD Publishing, pp. 79–93. https://doi.org/10.1787/9789264205390-en.

Fisher, W. R. (1985). 'The Narrative Paradigm: In the Beginning', *Journal of Communication*, 35(4), pp. 74–89. https://doi.org/10.1111/j.1460-2466.1985.tb02974.x.

Galera, J. and Baglioni, S. (2021). 'Dalla icercar di eroi alla costruzione di progetti comunitari. Perché è importante cambiare narrazione', *Impresa Sociale* 2/2021. Trento: Iris Network. http://doi.org/10.7425/IS.2021.02.10.

Jodha, N. S. (2000). 'Globalization and Fragile Mountain Environments', *Mountain Research and Development*, 20(4), pp. 296–299. http://doi.org/10.1659/0276-4741(2000)020[0296:GAFME]2.0.CO;2.

Klein, J. A., et al. (2019). 'Catalyzing Transformations to Sustainability in the World's Mountains', *Earth's Future*, 7, pp. 547–557. http://doi.org/10.1029/2018EF001024.

Machold, I. and Dax, T. (2017). 'Migration und Integration. Anstoß zur soziokulturellen Veränderung ländlicher Regionen Durch Internationale Migration', *Europa Regional*, 24, 3–4.

Membretti, A. (2021). 'Le Popolazioni Metromontane: Relazioni, Biografie, Bisogni', in Barbera, F. and De Rossi, A. (eds.), *Metromontagna. Un progetto per Riabitare l'Italia*. Roma: Donzelli.

Moulaert, F. and MacCallum, D. (2019). *Advanced Introduction to Social Innovation*. Cheltenham: Edward Elgar Publishing.

Pugh, R. and Dubois, A. (2021). 'Peripheries Within Economic Geography: Four "Problems" and the Road Ahead of Us', *Journal of Rural Studies*, 87, pp. 267–275. https://doi.org/10.1016/j.jrurstud.2021.09.007

Rappaport, J. (1995). 'Empowerment Meets Narrative: Listening to Stories and Creating Settings', *American Journal of Community Psychology*, 23(5), pp. 795–807. https://doi.org/10.1007/BF02506992.

Rodríguez-Pose, A. (2018). 'The Revenge of the Places That Don't Matter (and What to Do about It)', *Cambridge Journal of Regions, Economy and Society*, 11(1), pp. 189–209. https://doi.org/10.1093/cjres/rsx024.

Wittmayer, J. M., et al. (2019). 'Narratives of Change: How Social Innovation Initiatives Construct Societal Transformation', *Futures*, 112, p. 102433. http://doi.org/10.1016/j.futures.2019.06.005

Woods, M. (2005). *Rural Geography. Processes, Responses and Experiences in Rural Restructuring*. London: Sage Publications.

2.4 Thesis 4

International migration to rural and mountain areas is an important but neglected phenomenon

Ulf Hansson, Ingrid Machold, Thomas Dax and Per Olav Lund

Introduction

Over the past century, rural regions have been viewed as areas experiencing demographic decline. However, there has been a partial shift in migrants' destinations towards rural/mountain regions, and the increasing number of incoming migrants has induced manifold demographic and socioeconomic transformation processes in those areas (Kordel and Membretti, 2020). Since the 2000s, this phenomenon has gained more attention, both in national studies (e.g., Kasimis et al., 2010) and at a European level (Copus et al., 2021; Jentsch and Simard, 2009; McAreavey, 2017; Natale et al., 2019).

However, there is still the problem of the visibility and social recognition of foreign immigrants in these areas (Perlik and Membretti, 2018), probably because of their relatively small share of the population. Whilst in the whole of the EU, foreign-born residents account for 14.5% of the total population in cities, they represent 10.2% in towns and 5.5% in rural areas (Natale et al., 2019, p. 5). However, it is important to note that third-country nationals (henceforth TCNs) account for more than half of migrants in rural regions (Kordel and Membretti, 2020). The geographical distribution of these population movements is highly unequal, with positive net migration in large parts of Western European rural regions for more than two decades. Furthermore, the sheer numbers of TCNs (2.7 million in the EU in 2019, ibid.) make a crucial contribution to local economies and demography in these territories. Hence, the perception that foreign immigration to rural and mountain areas in Europe is a widely neglected phenomenon may also derive from corresponding narratives which describe rural areas as "neglected places" (see Thesis 3).

Previously, flows to mountain areas were conflated with the rise of destinations for second homes (with the focus on amenity migration, see e.g., Steinicke et al., 2012). More recent developments of migration to mountain areas concern a much wider range of types of migrants, including migrants

DOI: 10.4324/9781003260486-7

in need of protection. Concern for integration and efforts to establish appropriate "welcoming structures" has increased, at least in specific regions that place particular emphasis on achieving the smooth integration of, and the benefit from, the contributions of in-migrants (Gretter et al., 2017).

This chapter focuses on the widespread and increasing impact of international migration to peripheral areas of Europe. In 2015, with the enhanced refugee inflow to Europe, there was an explicit interest in the central and northern regions to distribute asylum seekers to rural areas in order to equalize the burden. This led to a balanced or even larger share of asylum seekers settling in rural areas (Proietti and Veneri, 2019, p. 172). In contrast, in the case of Spain, the ability to participate in the labour market was the main factor in dispersal and integration. By considering regional examples in Norway, Sweden, Austria and Spain, this chapter explores the extent to which awareness has been raised in contexts particularly affected by immigration and where activities to enhance integration are observed.

Sweden

Between 1970 and 1985, the importance of refugee and family migration was rising, and the shift to share burdens and improve social cohesion increased the number of refugees received by rural municipalities in Sweden (Kordel and Membretti, 2020). A study on immigration during the 1990–2010 period showed that net migration was, as in other Nordic countries, the main driver of population growth (Hedlund et al., 2017), and that Sweden issued the most residence permits in Nordic countries, mostly for asylum seekers and reunified family members (Karlsdottir et al., 2018). In sparsely populated areas and remote regions of Sweden, a positive international net migration was paralleled by a negative interregional net migration and a negative net fertility rate. Many of the small municipalities saw this as creating new opportunities, not only in terms of jobs in refugee reception facilities, but also as a basis for local services and increased tax revenues (Galera et al., 2018; Westholm, 2016). However, opportunities for employment were more numerous in urban areas (Vogiazides and Mondani, 2020), and as a result the lack of job opportunities for particularly forced migrants made them leave. Thus, using refugee reception as a strategy for municipal survival may only have limited potential, giving rise to what has been described as "social dumping." Once the flow of refugees has decreased, the municipalities receive reduced compensation. As a result, the surrounding infrastructure is likely to disappear, which may lead to a deterioration of services and employment opportunities. Herein lies a challenge for migrants and locals alike, because this may induce people to draw simplistic conclusions in which migrants are viewed as scapegoats (Mathisen and Stenbacka, 2015).

Mathisen and Stenbacka (2021) highlight aspects of tolerance and mutual respect regarding the reception of, and attitude towards, migrants/migration. They are grounded in multiculturalism and what has been described as a "generous nature" towards multiculturalism and newcomers' access to welfare. In light of developments in 2015 and 2016, the positive role played by civil society is important. Osanami Törngren et al. (2018) has pointed to the fact that government agencies were unable to cope with the situation in 2015 and 2016 without the support of civil society. For example, the Swedish Church played an active role by organizing language courses. Furthermore, the government allocated funding to civil society organizations in order to strengthen their support work for asylum seekers and newly arrived refugees (Arora-Jonsson and Larsson, 2021).

Norway

Immigration into Norway has increased in recent years. In 2021, 18.5% of the population were immigrants, or Norwegians born to immigrant parents, where approximately 11.5% were from non-EU27/EEC countries, and approximately 4.5% had a refugee background. Compared with other OECD countries, Norway is among those with the highest immigration rate relative to population size (OECD, 2018).

In 2003, Norway enacted an introductory programme intended to increase the opportunities for newly arrived immigrants (granted asylum seekers and UN transfer refugees and their reunified family members) to participate in working life, become financially independent and participate in society (Introduction Act, 2003).

Since the influx of refugees in 2015 and 2016, the political climate towards refugees has become more restrictive, and in the government's 2019–2022 integration strategy, resettlement is more targeted on education, qualifications, skills and the needs of the regional labour market (Lerfaldet et al., 2020). In 2021, part of the reform to tighten the immigration policy for asylum seekers and refugees, i.e., the Introduction Act, was replaced with the Integration Act, 2020. In the latter act, greater emphasis is placed on education, training and work, and expectations and responsibilities are clarified vis-à-vis regional/county authorities and the municipalities. As highlighted above in the case of Sweden, the settlement of migrants has been a strategic means with which to curb population decline and stagnation.

By cooperating with neighbouring municipalities on the services offered to immigrants, rural and remote municipalities can curb the negative consequences of the new resettlement policy. If they are able to improve the results of the introduction programme, rural and remote municipalities can still yield benefits from immigration, but this requires further efforts. Norway and its policies of integration have, on the one hand, aimed to ensure the access of minorities to equal social rights and opportunities for social mobility, while on the other they have encouraged diverse forms of cultural and religious identity, leading to debates and controversies concerning social participation as a precondition for integration. There is a perception that access to, for example, social security is increasingly conditioned (Djuve et al., 2017). Thus, integration is more and more linked to conditions which are supposed to strengthen opportunities for participation and inclusion.

Austria

International migration has led to a steady rise of the foreign population throughout Austria, particularly since the turn of the millennium: at present, about 15% of the inhabitants of Austria are foreign citizens and about 24% are of foreign origin (2020). The positive international migration balance has contributed to a persistent population increase in Austria, which also applies to the majority of its rural regions. These considerable immigration surpluses at least mitigate local population losses due to internal out-migration and ageing in most of these areas (Machold and Dax, 2017).

In 2015, immigrant numbers increased substantially due to the high number of asylum seekers, who accounted for almost 40% of all immigrants in that year (Statistics Austria, 2016, p. 41). Like other Federal States of the EU (e.g., Germany, Finland or Sweden), it was the explicit aim of the Austrian Federal State to establish accommodation facilities for asylum seekers in all regions. Through a dedicated legal act, the constitutional power of the federal state bypassed municipalities when establishing accommodation facilities, even if provinces, and districts of municipalities, opposed such plans (Rutz, 2017). There was a national agreement on an allocation formula of 1.5% asylum seekers in relation to total inhabitants of a municipality, which gave rise to an increased number of TCNs also in most rural parts of Austria.

In 2015, Austria's administration and civil society was alerted to migration and integration issues. Initially, public opinion was dominated by the acute needs of asylum seekers. Voluntary work and support concentrated on hosting and accommodating large numbers of people. To tackle the upcoming challenges better, in September 2015 a conference of Austrian mayors was held to discuss how to cope with the increasing number of refugees in Europe and to share good practices and experiences. Attitudes towards asylum seekers and refugees improved considerably in municipalities with fewer than 5,000 inhabitants (Bretschneider, 2016). Similarly, Schwabl (2015) highlighted that almost one in four Austrian inhabitants was engaged in refugee relief.

Despite this initially positive response, conservative beliefs and security concerns took over and impacted significantly on public discourse and politics, as exemplified by increasingly restrictive regulations. Furthermore, the legal employment of asylum seekers, which is only allowed under strict regulations, has been additionally restricted at the very local level, something that has intensified the impression among locals that asylum seekers are "lazy" (Machold et al., 2021). As a result of the strong centralization of regulations, this has had a hampering effect as activities for integration processes need also the consideration of the local and regional framework conditions.

Spain

In the sections prior, the focus has been on forced migration, and more specifically on refugees and asylum seekers. However, it is important within this context to flag up the experiences in Spain for the purposes of comparison. In Spain, the focus has been on labour migration. The economic and financial crisis of 2008 marked an important turning point in migration dynamics especially in Southern Europe. While in the case of Austria and Sweden a certain focus has been placed on forced migration, the case of Spain is slightly different in regard to the emphasis on labour migration. Spain, like Sweden, can be described as a supportive country in terms of access to services and equipment. Research, such as that by Aysa-Lastra and Cachón (2013), has highlighted the fact that the majority of immigrants may be able to improve their situation the longer they stay in the Spanish labour market.

With regard to integration, and similarly to Sweden, Spain also struggles to define an established integration model. Immigrants have found jobs that are not in demand by the native population, particularly in agriculture, construction and hospitality. Gaining employment equates these immigrants in relation to the social rights of citizens, guaranteeing a climate of remarkable coexistence.

As highlighted by the foregoing three case studies, a country like Spain is also vulnerable to economic crises that may halt this progress, since employment is considered an important factor in integration. Immigrants and their associations maintain a low profile in their social and political representation (Gobierno de Aragón, 2021).

International migration to rural and mountain areas can no longer be ignored

Multiple factors are at play in the context of migration and rural areas, regardless of the type of migration. In all the four case studies, there are challenges as well as opportunities. In the case of Sweden, challenges such as "social dumping" and "organizational fragmentation" regarding the administration of, for example, forced migrants, and short-term objectives in terms of state integration policies, highlight challenges on an institutional level. On a societal level, in 2015 there was a mobilization of civil society and a general welcoming attitude towards refugees in both Austria and Sweden. Since then, however, as highlighted above, the public discourse has shifted to a more restrictive one, and the perception of migrants – forced or labour – has changed, particularly in the case of Austria. Dispersal policies need to be accompanied by measures encompassing all dimensions of daily life if they are to be successful and long-lasting. Although civil society engagement since 2015 has made up for administrative reluctance, there is a particular need in mountain regions for political commitment at all levels if sustained integration is to be achieved.

Throughout these changes, the social diversity and attractiveness of mountain areas have remained an important driver of spatial movements. The argument that mountains (and other remote places) represent unique and highly valued assets which act as pull factors for new immigration groups has been taken up in local development strategies, like the LEADER program, LA21 or other good practice actions. As a matter of fact, also, COVID-19 restrictions have contributed to shifting national narratives towards valuing those places and providing inspiring examples of new activities in formerly marginal areas.

References

Arora-Jonsson, S. and Larsson, O. (2021). 'Lives in Limbo: Migrant Integration and Rural Governance in Sweden', *Journal of Rural Studies*, 82, pp. 19–28.

Aysa-Lastra, M. and Cachón, L. (2013). Segmented Occupational Mobility: The Case of Non-EU Immigrants in Spain. *Revista Española de Investigaciones Sociológicas*, 144: 23–47. doi: 10.5477/cis/reis.144.23.

Bretschneider, R. (2016). 'Flüchtlinge: Chancen für Gemeinden', *Record of Press Conference*, June 17, 2016. https://gemeindebund.at/website2020/wp-content/uploads/2020/07/Praesentation_Fluechtlingsstudie_Tischunterlage.pdf.

Copus, A., et al. (2021). 'European Shrinking Rural Areas: Key Messages for a Refreshed Long-term European Policy Vision', *TERRA. Revista de desarollo local*, 8, pp. 280–309. http://doi.org/10.7203/terra.8.20366.

Djuve, A. B., Kavli, H. C., Sterri, E. B. and Bråten. B. (2017). Introduksjosprogram og norskopplæring. Hva virker – for hvem? FAFO rapport 2017:31, Fafo, Oslo. https://www.fafo.no/images/pub/2017/20639.pdf.

Galera, G., et al. (2018). 'Integration of Migrants, Refugees and Asylum Seekers in Remote Areas with Declining Populations', *OECD Local Economic and Employment Development (LEED) Working Papers 2018/03*. Paris: OECD Publishing. Available at: www.foralps.eu/contenuti/allegati/84043b2a-en.pdf (Accessed 15 December 2021).

Gobierno de Aragon (2021). Plan Integral para la Gestión de la Diversidad Cultural de Aragón (2018–2021) – Comprehensive Plan for the Management of Cultural Diversity in Aragon (2018–2021). Zaragoza, Departamento de Ciudadanía y Derechos

Sociales, Dirección General de Igualdad y Familias, Gobierno de Aragón. https://www.aragon.es/documents/20127/674325/Plandiversidadcultural.pdf/187f41c2-b292-7df2-d640-3d4a892be7df.

Gretter, A., et al. (2017). 'Pathways of Immigration in the Alps and Carpathians: Social Innovation and the Creation of a Welcoming Culture', *Mountain Research and Development*, 37(4), pp. 396–405. http://doi.org/10.1659/MRD-JOURNAL-D-17-00031.1.

Hedlund, M., et al. (2017). 'Repopulating and Revitalising Rural Sweden? Re-examining Immigration as a Solution to Rural Decline', *The Geographical Journal*, 183(4), pp. 400–413.

Jentsch, B. and Simard, M. (eds.). (2009). *International Migration and Rural Areas. Cross-National Comparative Perspectives*. Studies in Migration and Diaspora. Ashgate: Farnham.

Karlsdottir, A., et al. (eds.). (2018). *State of the Nordic Region 2018: Immigration and Integration Edition*. Copenhagen: Nordregio, Nordic Council of Ministers and Nordic Welfare Centre. Available at: https://nordicwelfare.org/wp-content/uploads/2018/03/State-of-the-Nordic-Region-2018-Immigration.pdf (Accessed 15 December 2021).

Kasimis, C., Papadopoulos, A. G. and Pappas, K. (2010). 'Gaining from Rural Migrants: Migrant Employment Strategies and Socioeconomic Implications for Rural Labour Markets', *Sociologia Ruralis*, 50(3), pp. 258–276. http://doi.org/10.1111/j.1467-9523.2010.00515.x.

Kordel, S. and Membretti, A. (2020). 'Classification of MATILDE Regions. Spatial Specificities and Third Country Nationals Distribution', *Deliverable 2.1, H2020 Project Migration Impact Assessment to Enhance Integration and Local Development in European Rural and Mountain Regions (MATILDE)*. Bozen: EURAC.

Lerfaldet, H., et al. (2020). 'Anmodningskriterier for bosetting av flyktninger i 2019 [Request Criterias for Settlement of Refugees in 2019]', *Report 10/2020*. Bergen: Ideas2Evidence. Available at: www.ideas2evidence.com/sites/default/files/Anmodningskriterier%20for%20bosetting%20av%20flyktninger%20-%20ideas2evidence-rapport%2010_2020.pdf (Accessed 16 December 2021).

Machold, I. and Dax, T. (2017). 'Migration und Integration: Anstoß zur soziokulturellen Veränderung ländlicher Regionen durch internationale Migration', *Europa Regional*, 24(3–4), pp. 62–76.

Machold, I., et al. (2021). 'Austria', in Laine, J. (ed.), *10 Country Reports on Qualitative Impacts of TCNs* (MATILDE Deliverable 3.3), April 2021. http://doi.org/10.5281/zenodo.4726645.

Mathisen, T. and Stenbacka, S. (2015). Unge migranter skaper steder: Translokale og lokale praksiser i rurale områder i Norge og Sverige. In: Aure, M. et al. (eds.), *Med Sans for Sted: Nyere Teorier*. Bergen: Fagbokforlaget, pp. 213–229.

Mathisen, T. and Stenbacka, S. (2021). 'Sweden', in Laine, J. (ed.), *10 Country Reports on Qualitative Impacts of TCNs* (MATILDE Deliverable 3.3), April 2021. http://doi.org/10.5281/zenodo.4726645.

McAreavey, R. (2017). *New Immigration Destinations, Migrating to Rural and Peripheral Areas*. Abingdon: Routledge.

Natale, F., et al. (2019). *Migration in EU Rural Areas*. Luxembourg: Publications Office of the European Union. http://doi.org/10.2760/544298.

Osanami Törngren, S., Öberg, K. and Righard, E. (2018). 'The Role of Civil Society in the Integration of Newly Arrived Refugees in Sweden', in Lace, A. (ed.), *Newcome Integration in Europe: Best Pratices and Innovations Since 2015*. Brussels: FEPS, pp. 13–25.

Perlik, M. and Membretti, A. (2018). 'Migration by Necessity and by Force to Mountain Areas: An Opportunity for Social Innovation', *Mountain Research and Development*, 38(3), pp. 250–264.

Proietti, P. and Veneri, P. (2019). 'The Location of Hosted Asylum Seekers in OECD Regions and Cities', *Journal of Refugee Studies*, 34(1), pp. 1243–1268. http://doi.org/10.1093/jrs/fez001.

Rutz, J. (2017). *The Changing Influx of Asylum Seekers in 2014–2016: Austria's Responses, Short Summary of Study*. Vienna: European Migration Network.

Schwabl, T. (2015). *Österreich zwischen Hilfsbereitschaft und Fremdenfeindlichkeit, refugee report*. Baden: Marketagent.com.

Statistics Austria. (2016). *Migration und Integration 2016. Zahlen. Daten. Indikatoren 2016*. Vienna: Statistics Austria.

Steinicke, E., et al. (2012). 'In-migration as a New Process in Demographic Areas of the Alps: Ghost Towns vs. Amenity Settlements in the Alpine Border Area Between Italy and Slovenia', *Erdkunde*, 66(4), pp. 329–344.

The Introduction Act. (2003). *Lov om introduksjonsordning og norskopplæring for nyankomne innvandrere* (*introduksjonsloven*) (LOV-2003-07-04-80). Available at: https://lovdata.no/dokument/NLO/lov/2003-07-04-80 (Accessed 16 November 2021).

The Integration Act. (2020). *Lov om integrering gjennom opplæring, utdanning og arbeid (integreringsloven)* (LOV-2020-11-06-127). Available at: https://lovdata.no/dokument/NL/lov/2020-11-06-127 (Accessed 16 November 2021).

Vogiazides, L. and Mondani, H. (2020). 'A Geographical Path to Integration? Exploring the Interplay Between Regional Context and Labour Market Integration among Refugees in Sweden', *Journal of Ethnic and Migration Studies*, 46(1), pp. 23–45. http://doi.org/10.1080/1369183X.2019.1588717.

Westholm, E. (2016). 'Framtidstro och framtidsoro i flyktingfrågan', *Kungl. Skogs- och lantbruksakademiens tidskrift*, No. 5, pp. 8–11.

2.5 Thesis 5

Migration impact assessment as a powerful tool for evaluating the comprehensive effects of migration on local societies and economies

Birgit Aigner-Walder, Marika Gruber and Rahel Schomaker

Relevance of migration impact assessment

Migration Impact Assessment (MIA) evaluates the impacts of immigration in a broad, systematic economic and societal context (Nijkamp et al., 2012). In political and public debate, the focus is often on the challenges of migration and integration. Unless migrants are urgently needed on the labour market, worries of increased crowding-out processes, as well as concerns about the public costs of integration, fear of foreign infiltration or safety aspects dominate (OECD, 2011). Yet cultural diversity accompanied by immigration has great potential regarding innovativeness or creativity. There is a need to assess the combined effect of the aforementioned features of migration as a whole. Both, positive and critical, direct and indirect, impacts should be captured.

From a spatial point of view, MIA can be applied to nations, but also at a regional or local level, and it may reveal quite different impacts at different scales and among types of regions. The focus is on the advantages and disadvantages of migration and the attendant cultural diversity in the long run, and qualitative as well as quantitative techniques are applied. The effects of immigration on the host country or region may differ greatly – due not only to the heterogeneity of immigrants but also to differences among migration and integration policies or economic and societal circumstances in host countries or regions. MIA can be a powerful tool with which to enhance migrants' integration and local development. Moreover, when MIA comprises participatory tools, it is a means of empowerment and engagement of local communities and migrants.

By revealing shortcomings of existing literature and reflecting alternative, or supplementary methods for MIA, the thesis intends to increase the potential of comprehensive migration impact assessment in regard

DOI: 10.4324/9781003260486-8

to integration and community development issues in rural and mountain regions.

Standard measurement methods: insights and limitations

The scientific evaluation – as well as the public debate – in the context of migration tackles different dimensions of migration. On the one hand, economic as well as socioeconomic or social effects are distinguished, but also specific effects, e.g., in rural or urban areas, or more differentiated effects for specific social groups (e.g., low-skilled workers). The literature mostly focuses on the economic effects of labour migration, while the other forms of migration remain somewhat neglected, and so do the sociocultural aspects and territorial dimensions of migration (Caputo et al., 2021).

The methodological approaches of extant scientific studies comprise in particular single case studies, comparative case studies and quantitative econometric analyses using, e.g., different types of regression models or spatial models (Nijkamp and Mickiewicz, 2012; Bianchi et al., 2021). As for the indicators, different measures are used to capture the economic, fiscal and social dimensions of migration (see Table 2.1 below).

As regards the indicators used to date, several shortcomings are apparent. Firstly, many of the indicators delineated above are available on a national level only, and not on a regional or local one. Thus, analyses necessarily focus on the national level – a fact that may bias the respective outcomes because

Table 2.1 Standard MIA dimensions

Economic dimensions	*Fiscal dimension*	*Socioeconomic/ social dimensions*
Labour market (wage levels, employment rates, unemployment rates)	Costs for education or accommodation	Demography
Education, training, skills	Social benefits	Housing market
Productivity	Administrative expenses	Education system
Entrepreneurship	Tax contributions	Social infrastructure
(Social) Innovation	Social security contributions	Social mobility / social cohesion
Gross domestic product (GDP)		

Source: Own compilation

the impact of migration is significantly characterized by regional differences depending, e.g., on demographic developments or the urbanization of a region. Secondly, a problem of particular relevance to comparative research is the fact that comprehensive data for country comparisons are often missing because many countries do not collect statistics for all spatial levels or all dimensions of migration. Moreover, different measurement methods for indicators constitute a problem, e.g., regarding (un)employment rates. While these factors may be less significant in the European Union, they still exist and particularly hamper quantitative research based on comprehensive databases.

Moreover, several important aspects of migration effects, particularly important in rural regions, remain underresearched. Data problems as well as the focus on the above-described dimensions may be the reasons. While many "hard" facts, as delineated above, can be quantified relatively easy and are covered by databases, other, "soft" facts remain hidden. This applies particularly to effects discussed in the context of a "foundational economy" (see Thesis 9 of this Manifesto; also The Foundational Economy Collective, 2017). A foundational economy can be understood as comprising all public or private activities that provide the goods and services essential for the everyday lives of people, independent of the social status of single consumers. It is only partly covered by major databases. Many possible indicators are either more qualitative by nature or have not been collected by national statistics, e.g., when it comes to access to local social infrastructure, or food retail services in a specific area.

Especially in rural areas, migrants may have an important role with respect to local resilience and revitalization. Thus, important dimensions of the socioeconomic effects of migration remain obscure in the scholarly literature. Consequently, to capture the effects of migration on the foundational economy – e.g., social and health services, food production, energy, construction, retailers or tourism – new indicators and assessment methods should be taken into consideration.

Another issue that must be acknowledged is the fact that in recent years migration flows have become more diverse regarding the geographical and the sociocultural background(s) of migrants. While existing research primarily focuses on the educational or professional background of migrants, further studies may bring more sociocultural factors (e.g., social attitudes, cultural attitudes, ideals) to centre stage.

Advancing (participative) self-assessment/self-evaluation: relevance and benefits

An important method with which policymakers, administrative staff or NGO representatives can gain insights into immigration/emigration processes and the related intercultural coexistence of their inhabitants is self-assessment. In general, *assessment* means the "systematic collection, review, and use of information (about the respective object of study) for the purpose of quality improvement, planning, and decision-making" (Fredonia, 2021). Even though *assessment* is often used in the sense of "evaluation," the two words mean different things. *Evaluation* is not just assessment of the *status quo* of a situation; it also tries to measure the effects, impacts and unintended consequences of policies, programmes, strategies or measures so that policies and programmes effectively based on the results can be designed (Mertens and McLaughlin, 2004). OECD (2021) has defined six evaluation criteria which should help to guide evaluations and evaluative judgements: relevance, coherence, effectiveness, efficiency, impact and sustainability.

Evaluations are usually carried out in order to foster the accountability of an intervention (policy, measure, programme, etc.) or to facilitate learning. The results of the evaluation are used primarily to improve the intervention evaluated, to decide about its continuation or about the design of a new one. However, evaluation is not only about learning from experience, i.e., understanding something in retrospect and measuring changes ex post, or checking whether the intervention strategies used have had consequences. It also serves to control and steer complex social systems, as well as to counteract ongoing processes and enable changes when necessary. Overall, this notion of evaluation could foster a culture of learning (Batra et al., 2022).

In order to steer social development and to intervene with concrete (political) measures, the first requirement is fact-based knowledge creation to ensure the procedural character of learning and the constructive nature of knowledge (vs. an objective "fact") about the current situation in the municipality, district, province or federal state. Hence, an assessment of the current situation is the starting point. Within the H2020 MATILDE research project, different types of participatory evaluation methods are used: peer-to-peer exchange via local case study working groups and (policy) roundtables, or coevaluation approaches based on participatory action research methods and conducted via productive interaction between researchers and local partners. Moreover, a toolbox has been developed in order to foster the peer-to-peer approach as well, and to enable practitioners to assess and analyse the situation in their community by themselves. This activity can be called "self-assessment." If not only the current situation is to be assessed but also (political) measures implemented to transform this situation are to be evaluated for their effectiveness or impact, then "self-evaluation" is pertinent. However, a single perspective often cannot adequately capture the complexity of a certain social situation. In such cases, multiperspective participatory self-assessment/self-evaluation is advisable.

The foundation of participatory evaluation is Kurt Lewin's participatory action research theory. Lewin aimed to "raise the self-esteem of minority groups" (Adelman, 1993, p. 7), which should foster their "independence, equality, and co-operation" (Lewin, 1946, as cited by Adelman, 1993, p. 7). "The focus was transformative in nature and embraced concepts of empowerment, emancipation, liberation and self-determination, designed in large part to ensure that the least powerful would play a key part in the knowledge creation process" (Chouinard and Cousins, 2014, p. 6). Consequently, evaluation processes should be participatory, democratic and interactive in order to identify hidden points and include different stakeholder perspectives (Racino, 1999).

With the MATILDE Practitioner Toolbox, a step forward in data/information collection is carried out by the practitioners themselves (self-research) where possible, in collaboration with other experts participating in the participatory self-assessment/self-evaluation. Moderation of the process is carried out by a participating practitioner (who, in the case of more complex tasks, may take part in tailored trainings before or bring in an evaluation expert). The aim is to give as much self-responsibility as possible to the institutions concerned by empowering them through special tools, which should help them gain more knowledge about the current situation and start a process of participatory action development.

One example of participatory self-assessment is the development of a municipal profile. The aim is to gain an overview of the current and future demographic situation, the economy and labour market situation, the educational background of the population, the infrastructure (including, e.g., education and healthcare facilities or public spaces), the budgetary situation, the municipality's specificities and the social climate in the municipality. The procedure is based on quantitative and qualitative data. All stakeholders participating in the self-assessment contribute to the data provision, which is summarized by the moderator. During a round-table discussion, qualitative information, e.g., about the social climate in the municipality, is collected.

Participatory assessment/evaluation does not only include internal (e.g., programme staff) and/or external audiences (e.g., programme participants). If the evaluation is additionally designed to be inclusive and empowering by involving the participants in methodological decisions on the assessment/evaluation, power imbalances can be addressed and groups who often experience oppression and discrimination get the chance to express their opinions (Mertens and McLaughlin, 2004). "Empowerment Evaluation" does not only bring an innovative approach to evaluation; it can help to "create an environment conducive to the development of [people's] empowerment" (Fetterman and Wandersman, 2007, p. 182), which aims at "gaining control, obtaining resources, and understanding one's social environment" (Fetterman, 1995,

p. 2). Moreover, the participatory and a-hierarchical evaluation process creates space for learning and controversial debate and fosters the self-controlled management of evaluation results (Lucchini and Membretti, 2016). "Empowered communities" with their different types of stakeholders, such as policymakers, administrative staff, service providers and inhabitants, are able to promote equal knowledge exchange and production; they foster commitment to jointly agreed decisions and the resulting sustainable policies and services (Banks et al., 2013). Therefore, practitioners from the different disciplines as well as other civil society actors with and without migrant backgrounds become experts in their respective disciplines/everyday life.

Empowerment also strengthens minority people's agency, i.e., the "ability to take action or to choose what action to take" (Cambridge Dictionary, 2021), hence to "influence others, to negotiate, to affect change, and to make decisions" (Kwan and Walsh, 2018, p. 375). This ability is the key to self-evaluation. However, self-evaluation requires the actors involved to possess a high level of (self-)reflectivity. The degree of reflectiveness may differ among people, as well as among different local and national cultures. It presupposes cultures and governmental systems that allow and enable critical thinking, (self-)reflection, dissent and codetermination. This is a process that has to be learned (as early as possible). Different instruments to measure, e.g., intercultural competence or an organization's/individual's tolerance of ambiguity, are available (e.g., the Beliefs, Events, and Values Inventory; CILMAR, 2021).

Conceptual framework for interdisciplinary migration impact assessment

MIA should not be carried out by a single discipline or method. Only the integration of different fields of study enables a broadening of the knowledge base. Interdisciplinary research can lead to novel research aspects, to the development of new approaches or to furnish innovative insights within as well as among the participating disciplines (Morss et al., 2021). For this reason, an interdisciplinary migration impact assessment is indispensable, exemplarily focusing on economic, social and cultural dimensions.

Moreover, a combination of quantitative and qualitative assessment tools (mixed methods) yields instructive understanding of a research area. Furthermore, triangulation helps to validate research results. Quantitative and qualitative data should be collected at the same time and be treated as equally relevant to MIA. The results of the qualitative and the quantitative analysis are complementary, and they provide a comprehensive overview of the effects of migration. While quantitative data are used to analyse the socioeconomic and demographic situation of migrants (e.g., age, gender, labour force participation, education, income), qualitative approaches make it possible to gain in-depth information on the migrants' lives, satisfactions, fears, etc. An interdisciplinary approach can lead to new indicators developed across disciplinary boundaries.

Interdisciplinary research is generally marked by a strong need for coordination. Hence, for a successful interdisciplinary MIA, the following procedure is recommended:

- First, an interdisciplinary team must be created to carry out the MIA.
- Second, the framework for cooperation should be established; this includes (i) specification of the research questions and the research design, and (ii) explicit definitions of terms to avoid misunderstandings in the later research process.
- Third, the methods to be used should be defined, and indicators to be evaluated should be operationalized. In an interdisciplinary setting with mixed methods, coordination among researchers/disciplines is of great importance, because different concepts may be used to measure the same indicator.
- Start the collection of data with quantitative and qualitative methods simultaneously; interact regularly across disciplines to exchange ideas and perspectives.
- Assess the data within your discipline and discuss the results within the interdisciplinary team.

While the focus on the above-mentioned hard facts seems insufficient for an overall assessment of the effects of migration on the host country, the inclusion of "soft facts," e.g., indicators for foundational economy or sociocultural aspects, as well as other interdisciplinary indicators, seems of great importance.

Moreover, a MIA conducted solely by experts may underexpose relevant effects of migration. Self-evaluation in the context of self-assessment of the effects of migration, as well as migration and integration policies at local, regional or national level by politicians or civil servants themselves, may shed completely new light on the discussion and allow evidence-based decision-making.

Lastly, migrants as well as national inhabitants themselves should be included in the research process, because it is they that directly experience the consequences of successful / less successful integration.

References

Adelman, C. (1993). 'Kurt Lewin and the Origins of Action Research', *Educational Action Research*, 1(1), pp. 7–24. https://doi.org/10.1080/0965079930010102.

Banks, S., et al. (2013). 'Everyday Ethics in Community-based Participatory Research', *Contemporary Social Science*, 8(3), pp. 263–277. https://doi.org/10.1080/21582041.2013.769618.

Batra, G., Uitto, J. I. and Feinstein, O. (2022). *Environmental Evaluation and Global Development Institutions. A Case Study of the Global Environment Facility*. London and New York: Routledge.

Bianchi, M., Caputo, M. L., Lo Cascio, M., Baglioni, S., Aigner-Walder, B., Lobnig, C., Luger, A. and Schomaker, R. M. (eds.). (2021). *Comparative Report on TCNs Economic Impact and Entrepreneurship* (MATILDE Deliverable 4.4). https://doi.org/10.5281/zenodo.5017818.

Cambridge Dictionary. (2021). *Agency*. Available at: https://dictionary.cambridge.org/dictionary/english/agency (Accessed 17 December 2021).

Caputo, M. L., et al. (eds.). (2021). *10 Country Reports on Economic Impact* (MATILDE Deliverable 4.3). https://doi.org/10.5281/zenodo.5017813.

Chouinard, J. A. and Cousins, J. B. (2014). 'The Journey from Rhetoric to Reality: Participatory Evaluation in a Development Context', *Educational Assessment, Evaluation and Accountability*, 27, pp. 5–39. https://doi.org/10.1007/s11092-013-9184-8.

CILMAR – Center for Intercultural Learning, Mentorship, Assessment and Research. (2021). *Recommended Assessment Instruments*. Available at: www.purdue.edu/IPPU/CILMAR/Assessment/Instruments_We_Recommend.html (Accessed 17 December 2021).

Fetterman, D. (1995). *Empowerment Evaluation: A Form of Self-Evaluation*. San Francisco: American Educational Research Association. Available at: https://files.eric.ed.gov/fulltext/ED387517.pdf (Accessed 17 December 2021).

Fetterman, D. and Wandersman, A. (2007). 'Empowerment Evaluation. Yesterday, Today, and Tomorrow', *American Journal of Evaluation*, 28(2), pp. 179–198. https://doi.org/10.1177/1098214007301350.

The Foundational Economy Collective. (2017). *The Foundational Economy: The Infrastructure of Everyday Life*. Manchester: Manchester University Press.

Fredonia – State University of New York. (2021). *Assessment: Definition and Overview*. Available at: www.fredonia.edu/about/campus-assessment/assmtinforesources/overview (Accessed 9 November 2021).

Kwan, C. and Walsh, C. (2018). 'Ethical Issues in Conducting Community-Based Participatory Research: A Narrative Review of the Literature', *The Qualitative Report*, 23(2), pp. 369–386. https://doi.org/10.46743/2160-3715/2018.3331.

Lewin, K. (1946). Action Research and Minority Problems. In: Lewin G.W. (ed.) *Resolving SocialConflicts*. New York: Harper & Row.

Lucchini, F. and Membretti, A. (2016). 'The Land of Fires. Evaluating a State Law to Restore the Narrative Power of Local Communities', *PACO – Partecipazione e Conflitto*, 9(2), pp. 640–665. https://doi.org/10.1285/i20356609v9i2p640.

Mertens, D. M. and McLaughlin, J. A. (2004). *Research and Evaluation Methods in Special Education*. Thousand Oaks, CA: Corwin Press.

Morss, R. E., Lazrus, H. and Demuth, J. L. (2021). 'The "Inter" Within Interdisciplinary Research: Strategies for Building Integration Across Fields', *Risk Analysis*, 41(7), pp. 1152–1161. https://doi.org/10.1111/risa.13246.

Nijkamp, P. and Mickiewicz, A. (2012). 'Migration Impact Assessment: A Review of Evidence-Based Findings', *Review of Economic Analysis*, 4, pp. 179–208.

Nijkamp, P., Poot, J. and Sahin, M. (eds.). (2012). *Migration Impact Assessment*. Cheltenham, UK: Edward Elgar.

OECD. (2011). *Tackling the Policy Challenges of Migration: Regulation, Integration, Development*. Development Centre Studies. Paris: OECD Publishing.

OECD. (2021). *Applying Evaluation Criteria Thoughtfully*. Paris: OECD Publishing. https://doi.org/10.1787/543e84ed-en.

Racino, J. A. (1999). 'Qualitative Evaluation and Research: Toward Community Support to All', in Racino, J. A. (ed.), *Policy, Program Evaluation, and Research in Disability: Community Support for All*. New York, London, and Oxford: Haworth Press, pp. 3–22.

2.6 Thesis 6

Inclusion of migrants in rural and mountain territories is a multilevel and multidimensional process

Jussi P. Laine

Premise

The governance of human mobility, Mbembe (2019, p. 16) argues, is the most important challenge of the twenty-first century. Although written before the current COVID-19 pandemic, his argument still holds true, especially in terms of durability. Although surprisingly persistent, there are reasonable grounds to expect that the pandemic itself will one day be brought to an end, even if its impacts endure for a long time. In turn, human migration has been a persistent feature of human history and will continue to be so for years to come. For most of history, migration has been considered normal. Only recently have people on the move been depicted as a major problem. It is time to reverse this perception and, as Shah (2020) writes, turn migration from a perceived crisis into the solution for manifold socio-ecological changes. To follow Mbembe's (ibid.) argument, human mobility is normal; its governance is the challenge (see Thesis 7).

Formal laws, regulations and policies have been put in place in various countries to facilitate the settlement of immigrants and their integration into the host society. Various indicators have been defined to measure the level of integration (e.g., OECD/European Union, 2015), with the aid of which immigrants tend to be categorized in relation to their success in achieving the predefined integration benchmarks set against a normative framework and the presumably agreed, often nationally defined, standards. A glance at the prevailing public and political rhetoric suggests, first, that integration – understood as a sort of an ideal end state – is indeed a desirable, if unfeasible, goal and that success in this regard is still often considered to depend more on the immigrant's characteristics and actions than on those of the receiving society. *Society* here usually refers to the "country" into which immigrants are expected to incorporate themselves socioeconomically and adapt to its socio-cultural norms, values and customs. Used as a yardstick, integration thus continues to be assessed predominantly with quantitative measures of migrants'

DOI: 10.4324/9781003260486-9

socioeconomic performance, commonly in contrast with the "nonmigrant," "native" population (Ersanilli and Koopmans, 2011; Alba and Foner, 2016). Less attention is paid to who and what constitute the host population and society to which a migrant is expected to aspire. Central to the idea of a host population as a "norm" is that it consists of a homogeneous group into which immigrants should integrate (Saharso, 2019), that is, a *nation* state. It is argued here that this expectation is extremely biased and only accentuates the unfeasibility of the goal of integration by distorting the reality.

Integration – an end in itself?

Many of the dominant academic approaches continue to remind us that integration is a two-way process, in which both parties take an active part. Acknowledging this two-wayness has also become popular in the associated policy circles. However, while unquestionably valuable in terms of checks and balances, the mere realization itself does little to blur the social boundaries and binaries between "us" and "them," "insiders" and "outsiders," those who belong and those who are perceived not to belong. As Klarenbeek (2019) argues, the concept of two-way integration remains underdefined, and its mere endorsement is not enough to avoid or resolve the problems of one-wayness. As she convincingly shows, despite their good intentions, many dominant theoretical approaches to the two-way nature of integration have led to internal contradictions, only adding to and reinforcing, even if implicitly and unintentionally, a one-way integration discourse (ibid., p. 2). In other words, they have reconfirmed the existential separation between those who are considered to constitute "society" by default and those who do not, and who therefore need to "integrate" further (Schinkel, 2018).

Even if it is bidirectional, the process of change that the two parties are experiencing often remains completely unbalanced. This implies uneven power relations, normatively different responsibilities and thus different degrees of control over the process (Miller, 2016; Klarenbeek, 2019). In moving up the social ladder, being "well integrated" becomes the highest possible achievement for an outsider (Schinkel, 2013). Yet by this logic, even "well-integrated" immigrants can never truly become insiders, for if they could do so, they would not be "well integrated": integration would not even be an issue (Horner and Weber, 2011). As long as the unquestioned image of an "insider" continues to function as the benchmark against which the achievements of "outsiders" are compared, the reproduction of social boundaries, inequality and the perceived difference that they imply will continue to downplay any rhetorical commitments to the contrary. Such a reconfirmation of differences tends to divert attention from the deep-rooted structural factors that maintain inequality to individual ones perceived as beyond *our* control.

The broadly established and increasingly elaborate systems of monitoring, categorizing, and bordering reproduce otherness, maximizing its visibility in seeking to preserve and reconfirm a continuous positive version of the self, and the durability of the invisible social glue that is taken to hold "us" together. A "drawing self" is constantly present behind portraits of others (Chernobrov, 2016, p. 596). The more negative the qualities attributed to "them," the more positive "we" seem in comparison (Laine, 2020a, p. 75), and these representations seldom seek accuracy. Indeed, following Ahmed (2000, p. 19), strangers are not those we do not recognize but those we recognize as strangers. It is these subtle forms of sociocultural bordering to which attention must be paid, instead of fixating on integration in order to advance our societies' resilience, well-being, and fairness. Certainly, the immigrant's integration is not only shaped by explicit integration policies (Mügge and van der Haar, 2016); informal institutions among immigrants, e.g., religion and culture, can also determine its success, and immigrant groups may become either an accepted part of society on the same level as comparable native groups or they may isolate themselves or remain unrecognized and excluded (Garcés-Mascareñas and Penninx, 2016). However, more integration does not necessarily equate to more harmony, since changing power relations cause social friction (Klarenbeek, 2019, p. 13). An integrated society is not automatically "better" or more socially just.

Instead of continuing to fine-tune the analytical approach or fashioning yet another conceptualization of integration, this thesis claims that the entire premise from which the key postulation guiding our thinking stems needs to be re-evaluated. Like the nocturnal drunkard looking for his lost keys under a streetlight because that is where they are easiest to see, so have we, too, continued to seek solutions to the challenges that we are facing in the areas illuminated by our past state-oriented practices. Correspondingly, the key problems that European societies are addressing with their immigrant integration logics cannot be resolved through redefinitions or reappropriations of the term itself (Meissner and Heil, 2020). More fundamental readjustment is needed, and a key element of this endeavour is to advance the notions that people are not from a particular territorially bounded place and that the politics of belonging cannot be reduced to mere citizenship. Social, political and territorial demarcations persist largely because of the collective reproduction of their underlying essentialist conception and logic, yet belongingness and the conception of home are increasingly formed in a relationship to movement. In an increasingly mobile world, it is necessary to acknowledge the multilayer aspects of belonging, which often straddle the boundaries of nation and state.

The inclusion of migrants therefore needs to be understood as both a multilevel and multidimensional process, in which the different layers – if intertwined – may also have their own dynamics. While belongingness often still has to be enacted in the frame of the respective nation state and belonging to a certain nation continues to have an undeniable appeal (Laine, 2016, p. 471), the assumed deeply rooted ideal linkage between these concepts needs to be rethought – just as it is necessary to rethink the relationship between citizenship and the state. As Yuval-Davis (2006, p. 199) argues, people can simultaneously "belong" in many different ways and to many different objects of attachment. Belonging, she reasons, is not only about social locations and constructions of individual and collective identities and attachments but also about how they are valued and judged (ibid., p. 204). Belongings are seldom clear-cut; rather, they are fragmented and coincidentally rearticulated into constellations in which national identification plays only a part (Laine et al., 2020). This argument is put forward here in full acknowledgement of the incontrovertible fact that the idea of people being able to carry "bundles of rights" with them across nation state borders is one of the first lines of attack by nation states when a perceived "crisis" emerges (Collins, 2019).

To confront the simplistic relapse into state-centric thinking in times of "crisis" (Laine, 2020b), it is necessary to highlight the shifting power relations at the global, regional and local levels (García Andrade, 2018). Here, the increasing significance of the global level at the expense of regional integration and bilateral agreement has reawakened interest in interstate cooperation (Panizzon and van Riemsdijk, 2019). The rapid changes associated with globalization have decentred the state in some respects (Laine, 2016), yet the legal power to determine who is and who is not admitted to the territory of a particular country and recognized as a citizen is an enduring arena of state control (Weber and Tazreiter, 2021). European Union member-states have insisted on maintaining absolute control over "security matters" – a category into which migration is too often considered to fall. Despite the intra-EU levelling process intended to recommit member-states to the duty of solidarity, an *extra*-EU cooperation strategy for third countries has gained ground as the member-states have sought to renegotiate their obligations to the Union in order to maximize policy space around migration and asylum – that is, to reaffirm migration as "foreign policy" (Panizzon and Riemsdijk, 2019, p. 1228).

While numerous scholars have urged both the moral and economic imperative of migration be recognized in line with what Juss (2006) calls the "global public interest," pleas for unbounded inclusiveness (Laine, 2021) have fallen prey to renationalized populist politics and methodological nationalism. Within contexts of socioeconomic stress and geopolitical instability – of which a prime example is the current crisis inflicted by the persistent pandemic – a strong state tends to be offered and broadly accepted as a solution to the perceived chaos, and simplistic politics as a cure for the complexities that it has cultivated. These increasingly emotional reactions have not only largely overshadowed the scientific evidence pointing to the benefits of immigration, especially in rural and remote regions, but also allowed a blind eye to be turned to the increasingly evident notion, advocated by Jones and Wonders (2019; also Wonders and Jones, 2021), for example, that the policies of nation states often *produce* migration, harm and even violence.

Renegotiation, rejuvenation, resilience

While inclusive legal frameworks and efforts for effective global change are needed – the *Global Compact* being the first intergovernmentally negotiated agreement aimed at holistic and comprehensive change – the local level has been gaining importance as a setting for interethnic coexistence and participatory action. In contrasting the rise of crisis frames and a politics of exclusion, human agency deserves to be restored to the centre to redefine the meaning of belonging in a globalized world (Wonders and Jones, 2021). In this endeavour, the place-bound social relations of individuals offer great potential. Ultimately, despite globalization processes, place continues to be an object of strong attachment (Gustafson, 2014) and an inescapable aspect of people's everyday lives and its experiences (Butler and Sinclair, 2020, p. 64). However, once again, countering migrant exclusion requires more than universal appeals to inclusive egalitarian principles: abstract principles and norms need to be translated into concrete action, which is often more easily said than done. By illustrating opportunities associated with migration to rural areas and modes to revitalize them by alleviating the common challenges that many of these regions face, MATILDE's research aims to provide pragmatic paths to the kind of inclusivity envisaged by the Sustainable Development Goals and the long-term vision for the EU's rural areas adopted by the European Commission on June 30, 2021, which seek to make rural Europe more robust, connected, resilient and prosperous. Achieving inclusivity requires concrete knowledge and practical guidance that encourage local administrations and other stakeholders to see this as part of their responsibility and fulfil their role in this respect.

The rural and remote regions of Europe – regions often left behind – offer great potential not only for immigrants themselves and their respective new host societies but also, and above all, for something new to be created jointly and shared with others in the process itself of redefining "we." Placemaking by migrant populations has long been seen as an essential strategic response to the alienation, isolation and discrimination experienced by newcomers, because it helps cement new identities and sustain and empower marginalized communities (Phillips and Robinson, 2015, p. 414). However, these new places may be accompanied by negative implications of heightened ethnic difference through placemaking (Gill, 2010). It is argued here that a new understanding of being local, of belonging, should be sought through processes of inclusion and mutual recognition. These require continuous negotiation, but they fuel a social innovation in which the focus can be shifted from integration and assimilation to the cocreation of new transcultural spaces, economies and communities. This approach promotes social inclusion as nonlinear and reciprocal interaction through which new population groups negotiate new cultural meanings and concrete rights of citizenship with the existing populations, doing so within systems of socioeconomic, legal and cultural relations whose basic characteristics need to be considered if a sustainable, equitable and resilient society is to be created for all. The resulting communities will not only be different but will also be better adapted to thrive in the context of the current era's seemingly endless uncertainty.

References

Ahmed, S. (2000). *Strange Encounters: Embodied others in Post-coloniality*. London: Routledge.

Alba, R. and Foner, N. (2016). 'Integration's Challenges and Opportunities in the Wealthy West', *Journal of Ethnic and Migration Studies*, 42(1), pp. 3–22. http://doi.org/10.1080/1369183X.2015.1083770.

Butler, A. and. Sinclair, K. A. (2020). 'Place Matters: A Critical Review of Place Inquiry and Spatial Methods in Education Research', *Review of Research in Education*, 44(1), pp. 64–96. http://doi.org/10.3102/0091732X20903303.

Chernobrov, D. (2016). 'Ontological Security and Public (Mis)recognition of International Crises: Uncertainty, Political Imagining, and the Self', *Political Psychology*, 37(5), pp. 581–596. http://doi.org/10.1111/pops.12334.

Collins, J. (2019). 'Migration to Australia in Times of Crisis', in Menjivar, C., Ruiz, M. and Ness, I. (eds.), *The Handbook of Migration Crises*. Oxford: Oxford University Press, pp. 817–831.

Ersanilli, E. and Koopmans, R. (2011). 'Do Immigrant Integration Policies Matter? A Three-country Comparison among Turkish Immigrants', *West European Politics*, 34(2), pp. 208–234. http://doi.org/10.1080/01402382.2011.546568.

Garcés-Mascareñas, B. and Penninx, R. (2016). *Integration Processes and Policies in Europe – Contexts, Levels and Actors*. Dordrecht: Springer.

García Andrade, P. (2018). 'EU External Competences in the Field of Migration: How to Act Externally When Thinking Internally', *Common Market Law Review*, 55(1), pp. 157–200.

Gill, N. (2010). 'Pathologies of Migrant Place-making: The Case of Polish Migrants to the UK', *Environment and Planning A*, 42(5), pp. 1157–1173. http://doi.org/10.1068/a42219.

Gustafson, P. (2014). 'Place Attachment in Age of Mobility', in Manzo, L. C. and Devine-Wright, P. (eds.), *Place Attachment: Advances in Theory, Methods and Application*. London: Routledge, pp. 37–49.

Horner, K. and Weber, J.-J. (2011). 'Not Playing the Game: Shifting Patterns in the Discourse of Integration', *Journal of Language and Politics*, 10(2), pp. 139–159. http://doi.org/10.1075/jlp.10.2.01hor.

Jones, L. C. and Wonders, N. A. (2019). *Migration as a Social Movement*. Available at: www.law.ox.ac.uk/research-subject-groups/centre-criminology/centreborder-criminologies/blog/2019/05/migration-social (Accessed 17 November 2021).

Juss, S. S. (2006). *International Migration and Global Justice*. Aldershot: Ashgate.

Klarenbeek, L. M. (2019). 'Reconceptualising "Integration as a Two-way Process"', *Migration Studies*, 9(3), pp. 902–921. http://doi.org/10.1093/migration/mnz033.

Laine, J. (2016). 'The Multiscalar Production of Borders', *Geopolitics*, 21(3), pp. 465–482. http://doi.org/10.1080/14650045.2016.1195132.

Laine, J. (2020a). 'Ambiguous Bordering Practices at the EU's Edges', in Bissonnette, A. and Vallet, É. (eds.), *Borders and Border Walls: In-security, Symbolism, Vulnerabilities*. London, Routledge, pp. 69–87.

Laine, J. (2020b). 'Safe European Home – Where Did You Go? On Immigration, B/Ordered Self and The Territorial Home', in Laine, J., Moyo, I. and Nshimbi, C. C. (eds.), *Expanding Boundaries: Borders, Mobilities and the Future of Europe-Africa Relations*. London: Routledge, pp. 216–236.

Laine, J. (2021). 'Beyond Borders: Towards the Ethics of Unbounded Inclusiveness', *Journal for Borderlands Studies*, 36(5), pp. 745–763. http://doi.org/10.1080/08865655.2021.1924073.

Laine, J., Moyo, I. and Nshimbi, C. C. (2020). 'Borders as Sites of Encounter and Contestation', in Nshimbi, C. C., Moyo, I. and Laine, J. (eds.), *Borders, Socio-cultural Encounters and Contestations: Southern African Experiences in Global View*. London: Routledge, pp. 7–14.

Mbembe, A. (2019). 'Bodies as Borders', *From the European South*, 4, pp. 5–18.

Meissner, F. and Heil, T. (2020). 'Deromanticising Integration: On the Importance of Convivial Disintegration', *Migration Studies*, 9(3), pp. 740–758. http://doi.org/10.1093/migration/mnz056.

Miller, D. (2016). *Stranger in Our Midst. The Political Philosophy of Immigration*. Cambridge, MA: Harvard University Press.

Mügge, L. and van der Haar, M. (2016). 'Who is an Immigrant and Who Requires Integration? Categorizing in European Policies', in Garcés-Mascareñas, B. and Penninx, R. (eds.), *Integration Processes and Policies in Europe*. Cham: Springer, pp. 77–90.

OECD/European Union. (2015). *Indicators of Immigrant Integration 2015: Settling in*. Paris: OECD Publishing/Brussels: European Union. http://doi.org/10.1787/9789264234024-en.

Panizzon, M. and van Riemsdijk, M. (2019). 'Introduction to Special Issue: "Migration Governance in an Era of Large Movements: A Multi-level Approach"', *Journal of Ethnic and Migration Studies*, 45(8), pp. 1225–1241. http://doi.org/10.1080/1369183X.2018.1441600.

Phillips, D. and Robinson, D. (2015). 'Reflections on Migration, Community, and Place', *Population, Space and Place*, 21(5), pp. 409–420. http://doi.org/10.1002/psp.1911.

Saharso, S. (2019). 'Who Needs Integration? Debating a Central, Yet Increasingly Contested Concept in Migration Studies', *Comparative Migration Studies*, 7(16). http://doi.org/10.1186/s40878-019-0123-9.

Schinkel, W. (2013). 'The Imagination of "Society" in Measurements of Immigrant Integration', *Ethnic and Racial Studies*, 36(7), pp. 1142–1161. http://doi.org/10.1080/01419870.2013.783709.

Schinkel, W. (2018). 'Against "Immigrant Integration": For an End to Neocolonial Knowledge Production', *Comparative Migration Studies*, 6(31), pp. 1–17. http://doi.org/10.1186/s40878-018-0095-1.

Shah, S. (2020). *The Next Great Migration: The Beauty and Terror of Life on the Move*. New York: Bloomsbury Publishing.

Weber, L. and Tazreiter, C. (2021). 'Introduction: Migration and Global Justice', in Weber, L. and Tazreiter, C. (eds.), *Handbook of Migration and Global Justice*. Cheltenham: Edward Elgar, pp. 1–15.

Wonders, N. A. and Jones, L. C. (2021). 'Challenging the Borders of Difference and Inequality: Power in Migration as a Social Movement for Global Justice', in Weber, L. and Tazreiter, C. (eds.), *Handbook of Migration and Global Justice*. Cheltenham: Edward Elgar, pp. 296–313.

Yuval-Davis, N. (2006). 'Belonging and the Politics of Belonging', *Patterns of Prejudice*, 40(3), pp. 197–214. http://doi.org/10.1080/00313220600769331.

2.7 Thesis 7

International migration has to be considered as just one form among diverse mobilities

Tobias Weidinger and Stefan Kordel

Introduction

European rural and mountain areas were long considered to be areas of out-migration and demographic decline. Accordingly, discourses on "rural exodus" (Graham, 1892) or "rural flight" (Beetz, 2016) predominated in scientific and political debates. Immigration processes, both from abroad and from urban centres, in contrast, were often neglected despite the fact that they existed in areas in Europe considered to be peripheral, rural or mountainous, and despite the fact that immigration could be one solution for depopulation. When focusing on those regions today, one finds places that are characterized by a rapid change of (ethnic) diversity, while infrastructures and services able to adequately respond to it are lacking (Kordel and Weidinger, 2020c, p. 25). Hence, those regions could be classified as relatively novel destinations for immigrants – in other words, "new immigration destinations" (Winders, 2014; McAreavey, 2018; Kordel and Weidinger, 2020b, p. 507). Other places, instead, have already developed a certain migration history with continuous flows of immigrants that have led to a path dependency of immigration (Rodríguez-Pose and von Berlepsch, 2020). As a result of relationships forged during colonial times and spatial proximity to non-EU countries, specific migration regimes have been established and are continuously maintained, for example, between Spain and Latin American countries, between Italy and Spain and North African countries or between Scandinavian countries and Russia (Kordel and Weidinger, 2020b, pp. 507–508).

The remainder of the presentation of the thesis comprises an overview of the diversity of protagonists and processes of international immigration and domestic in-migration to new and established destinations in rural and mountain areas. To adequately understand the processes and resulting impacts on local development, however, one needs to consider the blurring and shifting boundaries of migration and mobility trajectories, as well as categorizations.

DOI: 10.4324/9781003260486-10

Rural mobilities and immobilities

In light of the increasing mobility of persons in Western societies, sociologists Mimi Sheller and John Urry (2006) criticized sedentarist assumptions about everyday lives and proclaimed a "new mobilities paradigm." Assuming mobility to be "normal," this paradigm had a considerable influence on migration scholarship. As a consequence, we must not consider migration as one single act but acknowledge ongoing negotiations of mobility and immobility (Halfacree and Rivera, 2012). Moreover, the paradigm also suggests a broader view on mobility, since migration processes only constitute a relatively small part of spatial movements, and blurring boundaries between residential mobilities and habitual/everyday mobilities are observable. For rural areas, the geographers Paul Milbourne and Lawrence Kitchen (2014) introduced the term *rural mobilities*, which encompasses "movements into, out of, within and through rural places; (…) linear flows between particular locations and more complex spatial patterns of movement (…) journeys of necessity and choice; economic and life-style based movements; hyper- and im-mobilities" (ibid., pp. 385–386). Current debates also stress unmarked categories of migration, e.g., they consider staying as an active process and a deliberate act (Schewel, 2019; see "rural staying," Stockdale and Haartsen, 2018) (Kordel and Weidinger, 2020c, p. 21). A distinctive feature of migration processes in rural and mountain areas today is its transient nature. Temporary movements, including cyclical, circular or seasonal mobilities, result in temporary presence of protagonists in rural and mountain areas. Protagonists may establish place attachments and belongings to those places but also to others, leading to the emergence of transregional or transnational social spaces (Glick-Schiller et al., 1992).

Human mobility to rural and mountain areas is also regularly described not only in terms of temporal and spatial characteristics but also with regard to the sociodemographic and socioeconomic profile, as well as the motivations of the protagonists. Regarding the latter, the distinction between voluntary and forced movements often predominates in public debates. Due to the cumulative causation of migration (Massey et al., 1998) and the fact that "both force and choice – structure and agency – are expressed within *all* migrations" (Barcus and Halfacree, 2018, p. 234), it is better to think of "voluntary" and "forced" as the two ends of a continuum (Weidinger, 2021, p. 16). Furthermore, "the interplay between agency and structure(s) often does not lead to migration from place A to a strictly defined destination B, but is better understood as a fragmented journey or trajectory" (Van der Velde and van Naerssen, 2011). To capture the complexity of migration histories of protagonists, "onward (im)mobilities" is regarded as more appropriate (Kordel and Weidinger, 2019).

New faces in rural and mountain places: a brief overview of processes of temporary/seasonal and permanent mobility

In what follows, we show the diversity of international immigration and domestic in-migration to rural and mountain areas, and we present the most important processes, i.e., humanitarian migration, student and labour migration, amenity/lifestyle migration and family migration.

Humanitarian migration

In accordance with allocation schemes and dispersal policies, many humanitarian migrants were and are mandatorily accommodated in European rural and mountain areas, at least for the duration of their asylum procedures, e.g., Austria, Finland, Germany, Italy, Norway, Sweden, Turkey and the UK (OECD, 2016). Other countries also assign a municipality or rural region to a humanitarian migrant after recognition of their status (e.g., Finland, Germany, Norway and Sweden, ibid.). Political justifications stem either from a policy of burden-sharing of costs and pressure on the housing market or from the assumption that small places are preferable sites for integration (Kordel and Weidinger, 2020a, p. 39).

Those third-country nationals who continue to stay in rural and mountain areas prefer small towns characterized by short distances. The reasons cited for doing so are a workplace or a language course for oneself and a (secondary) school for the children, established social ties and the friendliness of the local population. With regard to families, in particular, rural and mountain areas are constructed as safe places to raise children (ibid.). In everyday life, moreover, temporary absence and presence was and is common, for instance, to commute daily or weekly, or to go to cities to buy culturally appropriate food, visit friends and relatives or participate in religious feasts (Kordel and Weidinger, 2019). However, individual or household-related life events, such as a changing legal status, family reunification or the beginning of work, can result in a renegotiation of the residential location.

Student and labour. migration

Student migration is of minor quantitative importance. The temporary presence of guest students from a variety of countries is based on cooperative agreements of rural schools. Certain groups of university students may mandatorily move to rural and mountain areas on a temporary basis, e.g., to complete their clinical clerkships at rural hospitals or teacher-training courses at rural schools. Parallel to this, and based on the application of decentralization policies and the resulting instalment of higher education infrastructures, rural and mountain areas may also attract national and international students. In Germany, Scotland and Sweden, for instance, new universities were founded in rural and mountain areas in the past, whilst nowadays, research institutions and campuses of existing universities have branches in these areas. Per se, student migration is a temporal process of young individuals. Companies, municipalities and regions, however, endeavour to increase staying aspirations among high-skilled graduates and simultaneously prevent brain drain by smoothing the transition from university to work. As such, student migration is closely associated with labour migration, since residence permits may change once students graduate (Kordel and Weidinger, 2020a, pp. 41–42).

International and domestic labour migration is an important process in rural and mountain areas, especially for economic sectors characterized by a lack of workers, such as in mining, construction, industry, (health)care, domestic work, cleaning and hospitality. Agriculture, fishery and food processing are often entry points into the rural employment market and springboards to other sectors. Moreover, self-employment among newcomers is particularly important, e.g., in commerce or hospitality, where entry barriers are comparatively low.

Temporary or permanent labour migration from third countries is supported by specific visa regulations and bilateral labour arrangements, memoranda of understanding, as well as unilateral programmes that allow employers to recruit necessary workers (ibid., pp. 36–37). Apart from that, migrants also illegally overstay their tourist or study visas. Moreover, migration and cross-border commuting on a weekly, monthly or seasonal basis is facilitated by the freedom of movement in the European Union. Word of mouth and networks are often driving forces in identifying companies, while recruiting is supported by contractors and other intermediaries. Recently, rural municipalities and private organizations have started to engage in marketing efforts to attract qualified employees, e.g., doctors (ibid.). Nevertheless, the majority of migrants are not employed in knowledge-intensive sectors; instead, many of them work in poorly paid, temporary jobs, often in precarious circumstances. Some occupations are also highly gendered, e.g., men in construction or women in (health)care and domestic services; others, in turn, are strongly ethnicized. In the case of Turkey, for instance, Georgian men work in tea and hazelnut picking, Azerbaijanis gather fodder for animals and Afghan men work as shepherds (Dedeoğlu and Bayraktar, 2019).

Apart from permanent forms, migrant employment in rural and mountain areas is regularly characterized by short-term presences or marked by a distinct seasonality. In Italy, for instance, many migrants move across the country from one harvest to another or seek to find anticyclical jobs, which allow them to stay put in a place, while others return to the same company from abroad every year (Semprebon et al., 2017).

Labour migration can also be related to the existence of religious communities, military and school infrastructures or government agencies. International and domestic newcomers like priests, commanders, soldiers, teachers or state employees may (temporarily) be sent to rural and mountain areas on a no-choice basis by their institutions. Besides, permanent remigration to the place where oneself or one's parents grew up is a noticeable phenomenon, especially among middle-class migrants who have acquired their professional qualifications and plan to start a family. The motivation of migrants is a mixture of work-related aspects, lifestyle-related reasons or family needs, such as (anticipated) care of (grand)parents or takeovers of parental businesses. Some are also attracted by cheap building ground, inherited real estate or incentives aimed at young families. During the financial crisis in the mid-2000s, moreover, counter-urban movements arose especially in Southern Europe, whereby family support was an asset (Gkartzios, 2013). A similar trend has been apparent during the COVID-19 pandemic, facilitated by opportunities for remote working and the existence of digital infrastructures.

Amenity/lifestyle migration

Another group comprises consumption-led immigrants and in-migrants, i.e., amenity/lifestyle migrants. They are motivated to relocate to coastal regions (sea, lakes) and the hinterland, mountain villages, rural spa towns or the surroundings of metropolises both at home and abroad. Some of these migrants are induced to move temporarily or permanently by the availability of amenities such as high cultural and environmental quality or attractive landscapes and infrastructures ("amenity migration": see Moss and Glorioso, 2014). Others, instead, seek rural and mountain destinations in order to pursue a better way of life or to realize certain life goals ("lifestyle migration": see Benson and O'Reilly, 2009), often inspired by an idealized perception of these areas, i.e., the "rural idyll." Common to all of them is their relatively privileged socioeconomic, i.e., financial and time, resources, as well as legal status (e.g., freedom of movement, investment visa), which allow them to decide quite freely where they would like to live. Amenity/lifestyle migration can comprise different protagonists: (pre)retirees, middle-aged persons or young families and right-wing extremist settlers.

Amenity/lifestyle migrants purchase real estate to be used during weekends or holidays or as second homes or permanent places of residence. These purchases may be motivated by a certain cultural tradition, e.g., in the Czech Republic, Nordic countries or Russia, or by regular tourist stays in the destinations. Others instead are induced to buy property by differences in living and housing costs and good accessibility by car or low-cost carriers. Separation and divorce from, or the death of, a partner, loneliness and a lack of social contacts, as well as failure of one's life project, can result in reflection on the migration decision. Consequently, individuals may move on to other destinations (Kordel and Weidinger, 2019).

Family migrants

Family migrants are people who usually have no or few biographical ties to rural and mountain areas but who are closely related to the processes presented above. They move to such areas together with their partners or parents, or they join them months or years later. Family migrants are the partners and children of refugees, student migrants, labour migrants, return migrants, amenity/lifestyle migrants, as well as rural stayers. Regarding international migrants, family reunifications can be subject to certain requirements whereby relatives have to prove that they have the means of subsistence, sufficient housing space or a valid residence permit (Kordel and Weidinger, 2021).

Migration and mobility as new normality in rural and mountain areas

As was shown above, sociodemographic changes in rural and mountain areas are not limited to foreign migrants. Instead, international migrants are only one group of people among the many that, for various reasons, are currently on the move. It is also clear that the migration of diverse protagonists to rural and mountain destinations is not a one-time event, which includes place A and place B; rather, it is characterized by complex mobility pathways and results in the temporary and permanent connection of multiple localities.

Given the ubiquitous presence of residential and everyday mobilities and their intermediate forms, a new idea of mobility is proposed here: migration and mobility are the new normality in rural and mountain areas. For municipalities and practitioners on-site, it may be difficult to accept the transient decision-making processes of individuals and the temporary nature of residential location choices. However, migration and mobility, on the one hand, and temporary immobility on the other, should be regarded as a new norm, while communities must decide themselves whether they consider newcomers to be an additional burden that gives rise to fear or as a potential for local development (see also Thesis 5, Migration Impact Assessment). Regarding the latter, accordingly, communities should subordinate their actions and policies to the question of how to create an environment that motivates people to stay in rural and mountain areas not for the short run but the long one.

References

Barcus, H. R., and Halfacree, R. (2018). *An Introduction to Population Geographies. Lives Across Space*. London: Routledge.

Beetz, S. (2016). 'Der Landfluchtdiskurs – zum Umgang mit räumlichen Uneindeutigkeiten', *Informationen zur Raumentwicklung*, 2(2016), pp. 109–120.

Benson, M. and O'Reilly, K. (eds.). (2009). *Lifestyle Migration. Expectations, Aspirations and Experiences*. Farnham and Burlington: Ashgate.

Dedeoğlu, S. and Bayraktar, S. (2019). 'Refuged into Precarious Jobs: Syrian's Agricultural Work and Labor in Turkey', in Yılmaz, G., Karatepe, I. and Tören, T. (eds.), *Integration through Exploitation: Syrians in Turkey*. Augsburg and München: Rainer Hampp, pp. 68–79.

Gkartzios, M. (2013). '"Leaving Athens": Narratives of Counterurbanisation in Times of Crisis', *Journal of Rural Studies*, 32, pp. 158–167. http://doi.org/10.1016/j.jrurstud.2013.06.003.

Glick-Schiller, N., Basch, L. and Blanc-Szanton, C. (1992). 'Towards a Definition of Transnationalism. Introductory Remarks and Research Questions', *Annals of the New York Academy of Sciences*, 645(1), pp. 1–24. http://doi.org/10.1111/j.1749-6632-1992.tb33482.x.

Graham, P. A. (1892). *The Rural Exodus. The Problem of the Village and the Town*. London: Methuen & Co.

Halfacree, K. and Rivera, M. J. (2012). 'Moving to the Countryside…and Staying: Lives beyond Representations', *Sociologia Ruralis*, 52(1), pp. 92–114. http://doi.org/10.1111/j.1467-9523-2011.00556.x.

Kordel, S. and Weidinger, T. (2019). 'Onward (Im)mobilities: Conceptual Reflections and Empirical Findings from Lifestyle Migration Research and Refugee Studies', *Die Erde – Journal of the Geographical Society of Berlin*, 150(1), pp. 1–16. http://doi.org/10.12854/erde-2019-408.

Kordel, S. and Weidinger, T. (2020a). 'Migration Processes in European Rural and Mountain Areas', in Kordel, S. and Membretti, A. (eds.), *Classification of MATILDE Regions. Spatial Specificities and Third Country Nationals Distribution* (= Deliverable 2.1 of MATILDE project), pp. 31–47. http://doi.org/10.5281/zenodo.3999415.

Kordel, S. and Weidinger, T. (2020b). 'Patterns of Immigration of TCNs to MATILDE Countries and Regions in the Light of Wider Structural Transformations', in Kordel, S. and Membretti, A. (eds.), *Classification of MATILDE Regions. Spatial Specificities and Third Country Nationals Distribution* (= D2.1 of MATILDE project), pp. 508–514. http://doi.org/10.5281/zenodo.3999415.

Kordel, S. and Weidinger, T. (2020c). 'Conceptual Presuppositions on Migration and Place Attachment', in Kordel, S. and Membretti, A. (eds.), *Report on Conceptual Frameworks on Migration Processes and Local Development in Rural and*

Mountain Areas (= D2.4 of MATILDE project), pp. 19–25. http://doi.org/10.5281/zeonodo.4561788.

Kordel, S. and Weidinger, T. (2021). 'Germany', in Baglioni, S., Caputo M. L., Laine, J. and Membretti, A. (eds.), *The Impact of Social and Economic Policies on Migrants in Europe* (= D3.1 and D4.1 of MATILDE project), pp. 148–186. http://doi.org/10.5281/zenodo.4483950.

Massey, D. S., Arango, J., Hugo, G., Kouaouci, A. and Pellegrino, A. (1998). *Worlds in Motion: Understanding International Migration at the End of the Millennium*. Oxford: Clarendon Press.

McAreavey, R. (2018). *New Immigration Destinations. Migration to Rural and Peripheral Areas*. London: Routledge.

Milbourne, P. and Kitchen, L. (2014). 'Rural Mobilities. Connecting Movement and Fixity in Rural Places', *Journal of Rural Studies*, 34, pp. 326–336. http://doi.org/10.1016/j.jrurstud.2014.01.004.

Moss, L. A. G. and Glorioso, R. S. (eds.). (2014). *Global Amenity Migration. Transforming Rural Culture, Economy & Landscape*. Kaslo, BC: New Ecology Press.

OECD. (2016). *Making Integration Work. Refugees and Others in Need of Protection*. Paris: OECD Publishing.

Rodríguez-Pose, A. and von Berlepsch, V. (2020). 'Migration-prone and Migration-averse Places. Path Dependence in Long-term Migration to the US', *Applied Geography*, 116(3), Article 102157. http://doi.org/10.1016/j.apgeog.2020.102157.

Schewel, K. (2019). 'Understanding Immobility: Moving Beyond the Mobility Bias in Migration Studies', *International Migration Review*, 54(2), pp. 328–355. http://doi.org/10.1177/0197918319831952.

Semprebon, M., Marzorati, R. and Garrapa, A. (2017). 'Governing Agricultural Migrant Workers as an "Emergency": Converging Approaches in Northern and Southern Italian Rural Towns', *International Migration*, 55(6), pp. 200–215. http://doi.org/10.1111/imig.12390.

Sheller, M. and Urry, J. (2006). 'The New Mobilities Paradigm', *Environment and Planning A*, 38, pp. 207–226. http://doi.org/10.1068/a37268.

Stockdale, A. and Haartsen, T. (2018). 'Editorial Special Issue Putting Rural Stayers in the Spotlight', *Population, Space and Place*, 24(4), Article e2124. http://doi.org/10.1002/psp.2124.

Van der Velde, M. and van Naerssen, T. (2011). 'People, Borders, Trajectories: An Approach to Cross-border Mobility and Immobility in and to the European Union', *Area*, 43(2), pp. 218–224. http://doi.org/10.1111/j.1475-4762.2010.00974.x.

Weidinger, T. (2021). *Onward (Im)Mobilities and Integration Processes of Refugee Newcomers in Rural Bavaria, Germany*. Erlangen: FAU University Press.

Winders, J. (2014). 'New Immigrant Destinations in Global Context', *International Migration Review*, 18, pp. 149–179. http://doi.org/10.1111/imre.12140.

2.8 Thesis 8

Rural-urban relationships are a fundamental asset in terms of policies aimed at the inclusion of remote places within a metro-montane framework

Ayhan Kaya, Anna Krasteva and Susanne Stenbacka

Introduction

The spread of populist movements across Europe in recent years has been explained from a geographical perspective by changing political behaviour in rural, mountainous and remote areas rooted in socioeconomic differences, and with a theory of "revenge by places that do not matter." Rural-urban relationships consist of material and immaterial flows: people, economic resources, information, cultural and social capital, skills and practices. We shall focus on the drivers of populism in remote places, the politics of distraction formulated and implemented by various political actors, and the reterritorialization and revalorization of remote places in (post-)COVID and post-urban times.

Populism in remote places: socioeconomic, spatial and nostalgic deprivation

Populism is sometimes defined as a response to and rejection of the order imposed by neoliberal elites, an order that fails to use the resources of the democratic nation state to harness global processes for local needs and desires (Mouffe, 2018). Such populism originates in the deep-rooted structural inequalities and general impoverishment that mainstream political parties – on both the liberal right and the liberal centre-left – have actively contributed to in their embrace of neoliberal governance. In what follows, the focus will be on the dynamics of rural-urban relationships that constitute a fundamental asset of policies aimed at the inclusion of remote places by offering spatial justice within a metro-montane framework (Barbera and De Rossi, 2021).

DOI: 10.4324/9781003260486-11

In the past decade, especially since the global financial crisis hit the European continent as well as other regions of the world, one can observe different kinds of radicalization enacted by right-wing populists on the basis of antimulticulturalism, antidiversity, Islamophobia, antiglobalism, Euroscepticism and antivaccine prejudice. Right-wing populist parties and movements often exploit the issue of migration, especially the migration of Muslims, and portray it as a threat to the welfare and the social, cultural and even ethnic features of a nation (Ferrari, 2021). Populist leaders tend to blame a soft approach to migration for major problems in society, such as unemployment, violence, crime, insecurity, drug trafficking and human trafficking.

As Andrés Rodríguez-Pose (2018, pp. 196–198) put it:

> Populism as a political force has taken hold in many of the so-called spaces that do not matter, in numbers that are creating a systemic risk. As in developing countries, the rise of populism in the developed world is fuelled by political resentment and has a distinct geography. Populist votes have been heavily concentrated in territories that have suffered long-term declines and reflect an increasing urban/regional divide.

It is therefore not surprising that right-wing populism has become a recurrent phenomenon in remote places that "no longer matter" for the neoliberal political parties in the centre preoccupied with international trade, migration, foreign direct investment and urbanization. The perception of being "left behind" in remote places that "no longer matter" for the political centre may sometimes lead to what one might call *spatial deprivation*, or *spatial injustice*.

Citizens residing in remote places which "no longer matter" tend to be more attracted to antisystemic parties like right-wing populists because of their growing socioeconomic and spatial disadvantages. However, socioeconomic deprivation is not the only factor explaining populism's appeal. There are also some cultural factors that play an essential role. Many people nowadays experience what Gest et al. (2017) call "nostalgic deprivation," a term which refers to an existential feeling of loss triggered by the dissolution of established notions of identity, culture, nation, and heritage in the age of globalization. A growing number of people now crave job security, stability, belonging, a sense of future and also solidarity among workers, peasants and others. Similarly, those who live in remote, mountainous, rural places may also become dissidents against the neoliberal political centre (Droste, 2021). Those who have experienced long periods of decline, migration and brain drain, those that have seen better times and remember them with nostalgia and those that have been repeatedly told that the future lies elsewhere have used the ballot box as their weapon. Their sons and daughters are no different.

Those unable to go elsewhere for education or work have few options to find a compensatory form of control over their everyday lives apart from ethnonational radicalism, populism, nativism and sometimes White supremacism. Different forms of deprivation have been prevalent among the native youngsters who live in socioeconomically deprived remote places.

Populism, politics of distraction and rural-urban relations

The condition referred to above – i.e., rising populism in rural areas – is sometimes recognized and sometimes questioned. Whatever the case, it should also be understood in relation to parallel metropolitan problematics manifested in, for example, pronounced housing segregation, an increased concern for safety and security, marked poverty and vulnerability. In other words, associating populism with rural areas and emphasizing its connection to specific rural features may also be understood as a way for political holders of power to shift the focus from nearby urban shortcomings to unwanted processes in more remote, rural places.

This is not to deny that populism exists in rural areas; it is rather to question the sometimes-pronounced tendency to frame populism as intrinsic to rural regions or places, and indeed also to rural people. The concept of "distraction" can help shed light on this tendency. Distraction can be explained as a process whereby something prevents someone from concentrating on something else. One definition states that politics of distraction work in order to shift "the public's attention from the essential to the superficial" (Samwick, 2004, p. 5), where politicians are celebrities and politics become a form of entertainment. Such processes can take shape in any country or culture. A more extensive understanding concerns "a more or less conscious strategy pursued by those in politics who wish to accomplish their essential goals without excessive press scrutiny or any public awareness whatsoever" (Weiskel, 2005, p. 407). Of specific importance for the MATILDE Project, therefore, is an understanding of politics of distraction as involving a limited set of potential solutions to policy problems and where lived experiences are neglected, thereby diverting attention from complex structural forces.

Given the theme of this thesis, focusing on the message of rural populism as something that is attached to the rural population involves "moving the limelight from the close to the distanced" (Pred, 2000). The crisis of the urban and the inability to cope with urban problems thus constitute reasons for a reawakened interest in the situation and attitudes of rural people. In a work on racism with a focus on Swedish conditions, Pred (2000) argues for an understanding of the public debate that associates racism with the Swedish countryside – in parallel with failed integration processes in urban areas. He finds that at a time when a populism movement is gaining ground in

Sweden, specific rural places play leading roles in a debate and are labelled as if they are historically connected to right-wing movements.

The identified practice of locating racism and intolerance in specific places with excessive clarity can be seen as a way to mark the problem as concerning a very limited part of the population and the country. They are in fact processes that take place across regions and nations. Pred states that the stigmatization of such places rests on "*the misrepresentations of differences and the silencing of similarities*" (2000, p. 218). His point is thus that, with regard to openness towards newcomers as well as expressions of intolerance and racism, such approaches or attitudes are distributed *across* the country (albeit not evenly).

The separation of the rural and the urban is visualized in both spatial and temporal terms: geographical difference is accompanied by temporal difference; different places are separated by belonging to different temporal locations. Pred continues his analysis by stating that there is selective memory of selected places. Localizing, for example, Nazism or right-wing populism in a specific geographical area helps to hide the connection to other geographical areas. On the other hand, modernity – understood as educated, culturally developed and tolerant (people) – is closely connected to the urban, implying a discourse that rewards the urban over the rural. The urban is what may represent the modern nationalist self-image.

Distraction might thus be a tool for those who train the spotlight on suburban violence and crime and use this as an argument for equating migrants with criminals. On the other hand, it is a tool used to locate unwanted features, such as racism or intolerance, which are not considered to fit with the self-image of the modern human, in areas at a distance, a phenomenon also discussed in terms of "remotization" (see Thesis 1). From a rural standpoint, a recurrent description of an area as marginalized, and of its population as uneducated and afraid of change, leads to the invisibility of the real problems as well as "a spectacularly exaggerated denigration" (Ching and Creed, 1997, p. 4).

Rural-urban relationships are a fundamental asset for policies designed to create spatial cohesion, but these relationships may also work in a direction where spatial differences are used to explain larger political shortcomings and unwanted processes. A territorial stigma, arising from a communicative demarcation of a region or a place, has real consequences for those who feel the stigmatizing gaze focused upon them. To deal with such processes, one strategy might be to distinguish and demarcate spaces and groups of people (Meyer et al., 2016). A demarcation of international migrants is one example. Thus, we return to the issue of populism in relation to migration and to areas characterized by decline, loss of services and lack of faith in the future. The contemporary challenge is to turn the table and identify and communicate how migration may act as a vehicle to escape marginalization and a downward spiral. One way is to emphasize the opportunities that migration can bring, the role of migration as a means of connection where flows of people, cultures and resources can be sources of social innovation. In this endeavour, policymakers and public authorities are vital for evening out differences and enabling migration as a resource.

Local citizenship for the revalorization of remote places

The downward spiral trend of marginalization of remoteness can be reversed by the revitalization and revalorization of remote places through reterritorialization, migration, mobility, engagement, social innovation and citizenship. "Citizenship remains a significant site through which to develop a critique of pessimism about political possibilities" (Isin and Nielsen, 2014, p. 9).

Acts of citizenship

> To act…means to take an initiative, to begin (as the Greek word *archein*, 'to begin', 'to lead', and eventually 'to rule'), to set something in motion. The beginning is not the beginning of something, but of somebody who is a beginner himself.
>
> (Arendt, 1998, p. 177)

For Hannah Arendt, acting blends together agency, initiative, beginning, change – of the world and of the active self. This conceptual blend underlies the concept of citizenship as commitment, participation and transformation.

Volunteering and volunteers are acts and actors of local citizenship. The explosion of civic engagement during the migration crisis and the numerous intercultural initiatives in less dramatic times build the solidary citizenship, which impacts on both the local public sphere and the formation of actors-for-others. Solidary citizenship leads symbolic battles against the hegemonization of the populist discourses of b/ordering and othering and aims to transform the public space through the alternative discourses of solidarity, human security, human dignity and politics of friendship.

Refugees and migrant volunteers are the other figures of local citizenship. Translators, intercultural mediators, trainers of school football teams, participants in associations – of their community, but also various local organizations, etc. – and active migrants demonstrate the transformation of vulnerable individuals into empowered activists who assume participatory citizenship without/before legal citizenship and contribute to inclusive intersectionality and the transformation of the local public sphere.

The reterritorialization and revalorization of remote regions through migration and mobility in (post-)COVID, post-urban times

COVID-19 is an incentive for change in the symbolic attractiveness battle between remote and urban places. Digital nomads enjoy the advantages of "global" work combined with the charm of choosing oneself where to live. The local becomes more globalized; the global becomes more individually localized. The attractiveness battle started before the pandemic. New actors are entering the arena. Finns in a small Bulgarian village, already one-tenth of the local population, have transformed its image from a depopulated to an attractive place: several Bulgarians stopped selling their family homes, an IT expert created a village site to entice conationals to join the expat community. Sun and housing are the attractions for amenity migrants, who enjoy a mild climate, tranquillity, fresh air and cheap real estate. The symbolic impact is crucial – they change both the appearance and the image of remote places and rebalance the deterritorialization-reterritorialization nexus. Urban-rural mobility, the new ruralism in pre-, post-COVID times, is another phenomenon in the changing symbolic relationship between big cities and small towns and villages.

New actors of reterritorialization – migrants and mobile nationals – have started to counter the feeling of being "left behind" by revitalizing remote places, promoting the attractiveness of small settings and revalorizing remoteness (Gretter et al., 2017).

Diversity actors of social innovation in rural and urban settings

The actors of change are the vanguard of a politics of transformation that generates social innovations and – even more importantly – opens new venues for change. PlaySchool is an Oxbridge methodology introduced by a young British woman into a refugee centre on a Southern EU border. It is a triple educational and social innovation – in terms of methodology, crowdfunding and interculturality. Greening intersectionality is another emerging trend. The initiative called "Intercultural Gardens as Green Bridges" unites teachers and pupils, refugee and migrant children with children from minorities and the majority, mayors, migrants and locals, academics and NGO activists – all are equal and equally responsible for our shared life with nature.

From rural populism and politics of distraction to active citizenship

The global financial crisis, the "refugee crisis" and the COVID-19 pandemic have posed many challenges for citizens residing in both urban and rural spaces. There is an intense drain on social, cultural and economic capital in remote, rural areas, creating the so-called "spaces of emptiness." Those residing in remote places are likely to punish the political centres by aligning themselves with antisystemic, populist political parties. The politics of distraction has become prevalent in both rural and urban spaces. It has been revealed that the politics of distraction has become predominant because rural spaces are more often described by various political actors as marginalized, uneducated and afraid of change. It has also been argued that citizens residing in remote places generate active forms of citizenship that make it possible for them to reterritorialize and revalorize those remote regions through migration and mobility in (post-) COVID and post-urban times. Current developments such as the financial crisis, the "refugee crisis" and the pandemic make it clear that there should be stronger connections between urban and remote regions as far as policymaking processes are concerned. A metro-montane approach is needed to revalorize agricultural, mountainous and remote places distant from urban spaces.

References

Arendt, H. (1998). *The Human Condition*. Chicago and London: University of Chicago Press.

Barbera, F. and De Rossi, A. (eds.). (2021). *Metromontagna. Un progetto per Riabitare l'Italia*. Rome: Donzelli.

Ching, B. and Creed, G. (1997). *Knowing Your Place: Rural Identity and Cultural Hierarchy*. New York: Routledge.

Droste, L. (2021). 'Feeling Left Behind by Political Decision-makers: Anti-establishment Sentiment in Contemporary Democracies', *Politics and Governance*, 9(3), pp. 288–300.

Ferrari, D. (2021). 'Perceptions, Resentment, Economic Distress, and Support for Right-wing Populist Parties in Europe', *Politics and Governance*, 9(3), pp. 274–287.

Gest, J., Reny, T. and Mayer, J. (2017). 'Roots of the Radical Right: Nostalgic Deprivation in the United States and Britain', *Comparative Political Studies*, 51(13), pp. 1694–1719.

Gretter, A., Dax, T., Machold, I. and Membretti, A. (2017). 'Pathways of Immigration in the Alps and Carpathians: Social Innovation and the Creation of a Welcoming Culture', *MRD – Mountain Research and Development*, 37(4), pp. 396–405.

Isin, E. and Nielsen, G. (eds.). (2014). 'Introduction: Globalizing Citizenship Studies', in Isin, E. and Nyers, P. (eds.), *Routledge Handbook of Global Citizenship Studies*. London and New York: Routledge, pp. 1–11.

Meyer, F., Miggelbrink, J. and Schwarzenberg, T. (2016). 'Reflecting on the Margins: Socio-spatial Stigmatisation among Adolescents in a Peripheralised Region', *Comparative Population Studies*, 41(3–4), pp. 285–320.

Mouffe, C. (2018). *For a Left Populism*. Paris: Verso.

Pred, A. (2000). *Even in Sweden. Racisms, Racialized Spaces, and the Popular Geographical Imagination*. Berkeley, CA: University of California Press.

Rodriguez-Pose, A. (2018). 'The Revenge of the Places that Don't Matter (and What to Do about It)', *Cambridge Journal of Regions, Economy and Society*, 11, pp. 189–209.

Samwick, A. A. (2004). 'The Direct Line' [Editorial], *Rockefeller Center Newsletter*, 10(2), p. 5. Available at: http://rockefeller.dartmouth.edu/assets/pdf/newsletterf04.pdf.

Weiskel, T. C. (2005). 'From Sidekick to Sideshow – Celebrity, Entertainment, and the Politics of Distraction. Why Americans Are "Sleepwalking Toward the End of the Earth"', *American Behavioral Scientist*, 49(3), pp. 393–409.

2.9 Thesis 9
The social and economic development, attractiveness and collective well-being of remote, rural and mountain regions closely depend on the foundational economy

Filippo Barbera, Maria Luisa Caputo and Simone Baglioni

Becoming a citizen in rural, mountain and remote areas

MATILDE believes that the socioeconomic development and the well-being of rural and mountain regions is fostered by those areas that promote a practice of community belonging based on people's actual contribution to economic, social, cultural and political life instead of one grounded on only legal (e.g., based on formal citizenship) and normative (common origins or ancestry) assumptions of belonging. Hence, newcomers are considered to be part of the community when they perform "acts of citizenship," i.e., when, through their work provision, civic engagement, cultural sharing and appropriation, they take an active role in the provision, defence and reproduction of local commons.

A conceptualization of the economic inclusion of migrants which considers the "economy" not only as the usual sphere where the market, private interests and profit prevail but also as one which comprises all aspects related to the everyday needs of a community – from public service provision (like healthcare and education) to infrastructure (roads, communication networks, etc.) – is functional to such an understanding of community belonging and commons generation. This conceptualization of the "economy" goes under the name of "foundational economy." If we widen our notion of community belonging and economic inclusion practices to encompass "acts of citizenship," commons generation and the foundational economy, we can better assess the role that newcomers play in communities, and in particular in inner areas (e.g., areas that lie outside the "central" spatial focus of the

DOI: 10.4324/9781003260486-12

economic, social and political activities of a country), such as MATILDE remote and rural regions.

The foundational economy provides an insightful approach with which, on the one hand, to address the challenges of remote areas and, on the other hand, to appreciate how migrants contribute to the wider dimensions of a community life. In fact, the foundational economy comprises those sectors in which migrants play a key role in keeping remote and rural regions alive: it includes welfare services (social policy), grounded services (housing, utilities, health, education and care), mobility networks (transport systems), "palaces for the people" (libraries, community centres) and green infrastructures (parks, outdoor opportunities) (Collective for the Foundational Economy, 2018). These are the goods and services necessary for everyday life that are consumed by all citizens, whatever their wealth or income. Hence, if newcomers contribute to producing them and/or maintaining them economically and socially sustainable, they are better appreciated in society. Without such services, people's formal rights as citizens do not generate well-being and material freedom; consequently, newcomers work together with locals and sometimes on their behalf for such rights and well-being issues to be preserved. On the same ground, the foundational economy shelters those sectors of the economy that supply essential goods and services. At the general level, then, they are the material and the providential domains which are key to citizen entitlements (Gough, 2017). From this standpoint, these domains are of two main kinds. The first, the material one, comprises the infrastructures and services (pipes and cables, networks and branches) which connect households to daily essentials. These domains include the provision of water, electricity, retail banking and food. The second domain, the providential one, comprises crucial welfare activities like healthcare, education and basic income. Both domains require a long-term strategy and action, away from myopic policies towards long-term ones (see Krznaric, 2020), and able to take care of the interests of sometimes-voiceless subjects, such as migrants.

Furthermore, the importance of the foundational economy for the understanding of remoteness is connected to its spatially grounded nature: foundational economy goods and services are delivered to people through networks and branches which are spatially grounded (Schafran et al., 2021). This spatial significance may concern the between and within city level, as in Klinenberg's work, or regional disparities, as in the economic geography of inequalities (Rodríguez-Pose, 2017, 2020). In the following, we will illustrate how the foundational economy matters for a new conception of citizenship based on the active defence and management of local commons. To do so, we will first illustrate in some detail the connection between the foundational economy and rural, mountain and remote places.

Regrounding remote places

Since the financial crisis of 2007, regional disparities in growth and employment have widened, as confirmed by the Sixth Report on Economic, Social and Territorial Cohesion (CEC, 2014). In general, the dominant one-size-fits-all policies approach for lagging regions has not been the solution for regional development because it has proven unable to catch-up, less developed regions relative to metropolitan European regions (Tödtling and Trippl, 2005; Rodríguez-Pose and Ketterer, 2020). Besides the city level, territorial disparities are a key dimension for social infrastructure. Consider, for instance, the lack of formal social infrastructure that characterizes remote places, i.e., those areas which are far away from core citizenship services, such as schools, hospitals and public transport. The civic infrastructure of services like education, healthcare and transportation is framed as a precondition for a decent living, as much as it is for the creation of employment opportunities. This civic infrastructure is transscalar and is built through the enactment of social practices that unfold in daily life. In rural and remote communities, this civic infrastructure is described as "services of general interest" (SGI) and it is unlikely to prove economically viable, as the neoliberal ascendancy would maintain. SGI are first and foremost connected to the exploitation of citizenship rights and to the quality of life of (old and new) inhabitants. Accordingly, migration processes to rural regions – with the inflow of economic migrants – can contribute to alleviating those challenges. This is remarkable in the case of the departure of the working-age population and the "counter-urbanization" that involves movements from urban areas into accessible rural areas by people of retirement age, both of which phenomena are bound up with demographic ageing processes (see also Thesis 7 of this Manifesto on how international migration is one among the many types of mobilities from/to rural areas).

> Within a context of ongoing rationalisation and privatisation (…) the issue of service provision in remote and sparsely populated areas has thus become extremely problematic. Often the need to cut expenditures has coincided with increasing demands, due to an ageing population. Retirement migration also tends to place exceptionally heavy demands for health and care services on recipient areas. The provision of acceptable levels of public and private services in order to sustain adequate quality of life is one of the key policy challenges for rural areas.
>
> (Copus, 2011, p. 8)

In this context, migrants are important both for assuring service provision and for sustaining traditional fundamental economic sectors in remote regions (Natale et al., 2019; Bianchi et al., 2021), because they make it possible to respond to three critical needs. Firstly, they help to respond to the shortage of labour strictly related to an ageing population and to the out-migration of the working-age demographic classes, notably in traditional local foundational economic sectors, like agriculture or fishing, and in essential services like the care sector – as shown by MATILDE's assessment of migrants' economic impact (Caputo et al., 2021) as well as by earlier literature (Cangiano et al., 2009). Secondly, they contribute with their networks to the in-migration of young workers and young people, thereby counteracting the ageing and depopulation processes (Caputo et al., 2021). Thirdly, because of their different demographic profile (e.g., younger age and therefore a higher natality), they support the economic sustainability of the essential services and goods that constitute the civic infrastructure.

Poor accessibility to essential services due to marginalization is not the only feature that defines those territories. Low economic potential due to the distance from centres of economic activity, and a lack of relational proximity – understood as a disconnection with centres of power, which may discourage the active participation of local stakeholders in development policy – also define rural, remote and mountain regions (De Toni et al., 2020). In recent years, the concept of accessibility has shifted away from an economic perspective towards one centred on the quality of life, or the well-being of rural inhabitants (Noguera and Copus, 2016). In this sense, the marginality of these areas helps demonstrate the concept of spatial justice (Soja, 2010) with explicit consideration of space as an agent of social inequality reproduced by socioeconomic mechanisms that organize society *in* space. While urban areas generally provide essential services of adequate quality, rural areas are rich in unenhanced natural assets and cultural resources. They are rich in natural assets (water resources, agricultural systems, forests, natural landscapes) and cultural resources (archaeological sites, small museums, craft centres), and they have a complex territory shaped by diverse natural phenomena and human settlement processes.

To support territories as "living places," development strategies should not be space-neutral (Barca et al., 2012), because the geographical context matters in terms of its social, cultural and institutional features. Accordingly, place-based planning to support the way a grounded foundational economy constructs people-as-citizens requires formal institutions like local authorities and community organizations and initiatives to coordinate infrastructure provision, with schooling, business development, as well as the promotion of economic and social innovation. Policies such as these are directed towards reconstructing the nexus between local economic development and people's needs of social reproduction.

The foundational economy and performative citizenship

The foundational economy approach is rooted in a conception of citizenship rights and obligations that are derived from human needs. Needs are met through "intermediate satisfiers" that are context-dependent and responsive to sociotechnical system change – that is, they are social, cultural and place-specific, and they require agreement through continuous political negotiation and dialogue over time. The foundational economy thus requires citizenship to be considered as a localized set of practices geared to the management of common resources at a local level (Araral, 2014). Commons are characterized by the presence of a rival and nonexclusive common resource, a set of rules on access, withdrawal and uses of the resource shared by members of the commons (commoners), a regime of collective property or civic use (Ostrom, 1990). In this light, the foundational economy can be the basis for developing new practices of citizenship that revolve around the defence and management of local commons in rural, mountain and remote areas (Barbera et al., 2018; on rural commons, see Dalla Torre et al., 2021). Cases showing the specificities of the foundational economy in these areas are the so-called "community cooperatives." These are territorially bounded communities that share resources, work opportunities and services to improve a specific territorial context, binding this supply to a collective construction of the future. Although the phenomenon is still limited, community cooperatives (i.e., social cooperatives with a local focus) represent a key example of the defence and management of local commons in rural, mountain and remote areas (Bandini et al., 2014). In community cooperatives, the defence and management of local commons is directly linked to the guarantee of a range of services and goods able to satisfy the everyday needs of the community. To be a citizen means being involved in the social practices of defence and management of the local commons that constitute the backbone of daily life. Therefore, also, migrants who are not legally citizens of their country of settlement can claim that they belong to the local community. Community cooperatives are witnessing the participation of migrants that – through their active role and work – take care of local commons and thus "become" citizens in the eyes of others. Within this framework, community cooperatives are owned and managed by their members on the basis of inclusive principles: to be "one of us" requires actively participating in the reproduction of local commons. These principles are rooted in a community of people understood not only as residents on a given territory but also as a group of people who share interests, resources and projects for the well-being of that territory. The stakeholders are of different kinds (public, private for-profit, and nonprofit), and the production process involves the members of the local community as both producers and buyers. Thus, actions taken in defence of the local commons, as understood in light of the foundational economy, provide the basis for an inclusive local community. This is not confined to community cooperatives, for it includes other forms of cooperative organizations. A case in point is "SIPA," which was established in 1994 in South Tyrol (Bolzano) as a "type B" social cooperative for the inclusion of disadvantaged people in accordance with the Italian Law 381/91 on social cooperatives. SIPA is mainly active in the sector of cleaning services and collective catering, and it has a range of action covering the entire South Tyrol. Its clients are principally retirement homes for elderly people, hospitals, offices and public administrations. The cooperative has 110 members, of whom 100 are working members, and 62 of them are foreigners. The annual budget of the company is 1.5 million euros. Another example is "Il Frutto Permesso," an agricultural cooperative founded in 1987 and located in Bibiana (Pellice Valley), within the territory of the city of Turin, in a predominantly mountainous area characterized by a mixed economy of which agriculture accounts for 14%. In this area there is a prevalence of livestock raising but also a component of agricultural production, especially fruit growing. The cooperative was one of the first in Piedmont to convert its activity entirely to organic farming. It produces fruit and vegetables, as well as fodder for animals, with a client base covering the whole metropolitan area of Turin, and an annual budget of 2 million euros. The company has 25 permanent workers, 10 of whom are foreigners (40%). During harvest time, foreigners account for 50% of the manpower, including several temporary workers.

Overall, through the lenses of the foundational economy, people belonging to local communities stand in stark contrast to the neoliberal notion of migrants' active citizenship, a prescriptive and depoliticized "tick-box" exercise based on law-abidingness and individual responsibilization, as in the case of the British naturalization test introduced after the civil disturbances in the northern towns of Oldham, Burnley and Bradford in 2001 (Bassel et al., 2021). Yet migrants' participation in local commons may be limited by their legal status, and it may also be the subject of a debate between a universalistic access to local resources versus a communitarian one. Hence, the arrival of new populations in rural and remote areas prompts rediscussion of the ownership of common goods, such as land, water, landscape and local knowledge (Membretti and Viazzo, 2017), and a review of the boundaries of the "local community." Nevertheless, by being actors of citizenship and therefore actively participating in the life and management of local commons, migrants and local communities work towards building a shared community vision and strive for its realization.

References

Araral, E. (2014). 'Ostrom, Hardin and the Commons: A Critical Appreciation and a Revisionist View', *Interrogating the Commons*, 36, pp. 11–23.

Bandini, F., Medei, R. and Travaglini, C. (2014). 'Community-Based Enterprises in Italy: Definition and Governance Models', *SSRN Electronic Journal* [Preprint]. Available at: https://ssrn.com/abstract=2408335.

Barbera, F., Negri, N. and Salento, A. (2018). 'From Individual Choice to Collective Voice. Foundational Economy, Local Commons and Citizenship', *Rassegna Italiana di Sociologia*, LIX(2), pp. 371–397.

Barca, F., McCann, P. and Rodríguez-Pose, A. (2012). 'The Case for Regional Development Intervention: Place-based Versus Place-neutral Approaches', *Journal of Regional Science*, 52(1), pp. 134–152.

Bassel, L., Monforte, P. and Khan, K. (2021). 'Becoming an Active Citizen: The UK Citizenship Test', *Ethnicities*, 21(2), pp. 311–332.

Bianchi, M., *et al.* (2021). *Comparative Report on TCNS' Economic Impact and Entrepreneurship. Deliverable 4.4.* https://matilde-migration.eu/wp-content/uploads/2021/08/d44-comparative-report-on-tcns-economic-impact-and-entrepreneurship.pdf

Cangiano, A., *et al.* (2009). *Migrant Care Workers in Ageing Societies: Research Findings in the United Kingdom.* Oxford: COMPAS University of Oxford.

Caputo, M. L., et al. (2021). *10 Country Reports on Economic Impacts. MATILDE Deliverable 4.3.* https://matilde-migration.eu/wp-content/uploads/2021/07/d43-10-country-reports-on-economic-impacts.pdf.

CEC (2014). *Sixth Report on Economic, Social and Territorial Cohesion.* https://ec.europa.eu/regional_policy/sources/docoffic/official/reports/cohesion6/6cr_en.pdf

Copus, A. (ed.). (2011). *EDORA. European Development Opportunities for Rural Areas. Applied Research 2013/1/2. Final Report.* Luxembourg: ESPON & UHI Millennium Institute, 2011 (ESPON).

Dalla Torre, C., Gretter, A., Membretti, A., Omizzolo, A. and Ravazzoli, E. (2021). 'Questioning Mountain Rural Commons in Changing Alpine Regions. An Exploratory Study in Trentino, Italy/Aprire il dibattito sui commons rurali di montagna nelle regioni alpine in cambiamento. Uno studio esplorativo in Trentino, Italia', *Revue de Géographie Alpine*. http://doi.org/10.4000/rga.8589.

De Toni, A., Di Martino, P. and Dax, T. (2021). 'Location Matters. Are Science and Policy Arenas Facing the Inner Peripheries Challenges in EU?', *Land Use Policy*, 100, p. 105111.

Gough, I. (2017). *Heat, Greed and Human Need.* Cheltenham: Edward Elgar Publishing.

Krznaric, R. (2020). *The Good Ancestor: How to Think Long Term in a Short Term World.* London: WH Allen.

Membretti, A. and Viazzo, P. P. (2017). 'Negotiating the Mountains. Foreign Immigration and Cultural Change in the Italian Alps', *Martor*, 22, pp. 93–107.

Natale, F., *et al.* (2019). *Migration in EU Rural Areas.* Luxembourg: Publications Office of the European Union. Available at: https://publications.jrc.ec.europa.eu/repository/handle/JRC116919.

Noguera, J. and Copus, A. (2016). 'Inner Peripheries: What Are they? What Policies Do They Need?', *Agriregionieuropa*, Anno 12(45).

Ostrom, E. (1990). *Governing the Commons: The Evolutions of Institutions for Collective Actions*. New York: Cambridge University Press.

Rodriguez-Pose, A. (2017). 'The Revenge of the Places that Don't Matter (and What to Do about It)', *Cambridge Journal of Regions, Economy and Society*, XI(1), pp. 189–209.

Rodriguez-Pose, A. and Ketterer, T. (2020). 'Institutional Change and the Development of Lagging Regions in Europe', *Regional Studies*, 54(7), pp. 974–986. http://doi.org/10.1080/00343404.2019.1608356.

Schafran, A., Smith, M. N. and Hall, S. (2021). 'Agency, Capabilities and Geographical Politics: A Book Review Symposium', *Progress in Human Geography*, 45(6), pp. 1731–1740.

Soja, E. W. (2010). *Seeking Spatial Justice*. Minneapolis: University of Minnesota Press.

Tödtling, F. and Trippl, M. (2005). 'One Size Fits All?: Towards a Differentiated Regional Innovation Policy Approach', *Regionalization of Innovation Policy*, 34(8), pp. 1203–1219.

2.10 Thesis 10

The COVID-19 pandemic: threats and opportunities for remote, rural and mountain regions of Europe, and for their inhabitants

Marika Gruber, Nuria del Olmo-Vicén and Raúl Lardiés-Bosque

Introduction

Rural areas and their populations, in particular migrants, have been especially hard-hit by the COVID-19 pandemic due to their fragile and precarious living conditions. A recent study conducted by the International Organization of Migration (Guadagno, 2020) analysed the specific ways in which migrants have been affected by the pandemic. It has identified a variety of conditions which make migrants more vulnerable in times of such a pandemic: limited awareness of recommended prevention measures (also due to linguistic barriers); limited right to receive healthcare; inability to respect social distancing because of crowded, multigenerational homes; reliance on public transportation; employment in close-contact professions; limited access to hygiene items (Guadagno, 2020). Liem et al. (2020) identified international migrant workers as a special vulnerable group and suggested measures such as public health campaigns in multiple languages to protect migrants and help them receive adequate healthcare, as well as avoid community infection.

Owing to natural demographic development but also because of the low pay and bad prestige of some jobs, there is also a general shortage of routine or skilled workers for certain sectors in some rural regions, which is partly, or sometimes largely (as in agriculture or tourism), compensated by migrants. The COVID-19 pandemic has highlighted the dependence on migrant labour of these sectors, a dependence which became especially evident with the closure of borders. The border closures caused by the COVID-19 crisis have had various impacts on rural areas and the immigrant populations living in them. Firstly, COVID-19 is generating a stagnation

DOI: 10.4324/9781003260486-13

of migratory flows (Ardis and Laczko, 2020) that impacts on the hiring of temporary workers in rural areas, but also the arrival of refugees. Secondly, the closure of borders also makes return migration to countries of origin impossible (Landis MacKellar, 2020).

However, it should not be concluded that rural areas in general provide more difficult conditions, in particular for migrants. While metropolitan cities and urban areas have become hotspots of the pandemic, followed by lockdowns all over the world and high numbers of COVID-19 victims, rural areas have received (especially during the first COVID-19 wave in 2020) increased attention as places of higher safety and liveability and as producers of regional food and utility items in times of border closures (Dettling, 2020; Membretti, 2020).

This chapter discusses the challenges to rural areas and their immigrant population caused by COVID-19, as well as the advantages generated during and after the waves of the pandemic. It outlines new pull factors that may enhance the attractiveness of rural areas to migrants, as well as socioeconomic and environmental advantages that have increased or even emerged for these areas in times of (post)-COVID-19. The chapter also makes some policy recommendations which should help to attract new settlers and immigrants to rural areas in Europe.

Impact of COVID-19 on immigrants in rural areas

The pandemic has highlighted certain characteristics of rural and mountain areas that have impacted, both positively and negatively, on their inhabitants and, particularly, on immigrants (Gruber et al., 2021). The main impacts on labour and training activity, on health and on the processes of integration and social cohesion are briefly outlined in what follows.

Negative impacts

Vulnerable populations are doubly affected by the crisis. First, because they are often more at risk from a health standpoint. Second, because they are particularly hard-hit by the economic crisis (OECD, 2020b). Indeed, the COVID-19 pandemic has had particularly negative socioeconomic impacts on migrants residing in rural and mountain areas, and particularly on those who are unskilled workers.

The reduction (or stagnation) of sectors of activity typical of rural and mountain areas – rural tourism and winter sports – has had a significant impact because they usually employ a significant percentage of foreigners (Gruber et al., 2021). In fact, the COVID-19 crisis has not only caused a loss of jobs, especially in the services sector, but has also permanently reduced the number of vacancies and complicated the application process (Machold et al.: "Austria," in Laine, 2021). Moreover, at the beginning of the pandemic in 2020, the closure of administrative services prevented or slowed down the renewal of residence and work permits, thereby reducing job opportunities for foreigners.

Regarding health risk, the impossibility of teleworking increases the risk of infections for workers in the agricultural and food processing sectors, which also employ a larger number of immigrants in rural European areas, as shown by studies on the impact of COVID-19 based on territorial differences (OECD, 2021). The risk has especially increased for women immigrants working in elderly care activity, because of the high rates of older people living in rural areas and who have been widely affected by COVID-19. Moreover, the most vulnerable people have been those with irregular jobs, who, lacking a contract, have not been able to justify the commute for work reasons.

The COVID-19 pandemic has also impacted on the training and development of human capital, particularly in linguistic immersion and specialized professional training, which may have a negative impact on job placement. The increase of online training has also constituted an important barrier for TCNs (Caputo and Baglioni: "United Kingdom," in Laine, 2021).

Finally, the health alert has had a detrimental impact on social integration and cohesion due to the decrease in personal interactions between locals and newcomers, which in some rural regions were already difficult before COVID-19 (Gruber et al.: "Austria", in Caputo et al., 2021). Moreover, COVID-19 measures have increased physical distances between people, nor does the use of masks help, because they intensify social and intercultural distance.

Positive impacts

On the other hand, it is important to highlight that the pandemic has also had positive effects in rural and mountain areas for all their inhabitants, both indigenous and immigrant.

Firstly, the pandemic has demonstrated the constant need for "essential" workers in sectors that provide crucial services – such as healthcare – and that keep supply chains running (e.g., agriculture) (OECD, 2021; Papademetriou and Hooper, 2020). Moreover, the difficulty of hiring new and/or receiving foreign seasonal workers due to the closure of borders has favoured migrants residing in rural areas, making them less vulnerable than their compatriots residing in urban areas.

Secondly, the increase in the number of people moving from urban to rural areas after the lockdown during 2020 has been one of the most significant processes. The influx of other residents has increased the income of retail stores in rural and mountain areas. Similarly, the impossibility of going to urban areas has forced locals to buy directly from the closest small retail stores (Bianchi et al., 2021); these include small businesses owned by immigrants who have found the opportunity to expand their products and services beyond their cultural group. Critical conditions have also triggered innovation in rural and mountainous areas, accelerating the digitization process, so that these services can now more easily overcome the barrier of remoteness and access wider markets (Gruber et al.: "Austria," in Caputo et al., 2021). For this reason, the COVID pandemic can also be seen as an opportunity that has enabled many immigrants to develop or learn digital skills.

The various waves of the pandemic have highlighted the complex challenge that migration management represents. On the one hand, for the most recent unskilled immigrants (mostly TCNs with a less stable legal administrative situation, like Latin American women caregivers in Southern European countries or others working in bars and restaurants), the pandemic has limited their work activity and mobility. Nevertheless, this has also evidenced the need for these essential workers and placed them at the centre of policies intended to improve their access to services, their legal administrative situation and their training.

On the other hand, immigrants with longer settlement in the country (with permanent residence and work permits) and highest training (especially those with jobs in the knowledge society) have been able to leave the urban centres to move to rural and mountain areas.

In any case, the negative impacts can be transformed into positive ones through appropriate policies, as now described.

Resilient strategies to cope with the COVID crisis in rural and mountain areas

As stated above, foreign immigrants living in rural areas have been among the most vulnerable and affected groups during the pandemic. To deal with this situation, national, regional and also local governments have approved several economic and labour measures since the onset of the pandemic. The measures identified in this section are selected examples taken from different countries participating in the MATILDE Project.

The agricultural sector is considered to be an essential sector in many countries. Hence, many measures were enacted in 2020 and 2021 to ensure the collection of primary products and the hiring of seasonal immigrants for the fruit picking (European Commission, 2020). On 30 March 2020, the European Commission asked member-states to guarantee and facilitate the movement of seasonal workers in order to avoid serious labour shortages in production sectors characterized by seasonality, expressly including agricultural. As a consequence of the shortage of these workers, many European countries opted for the partial regularization of foreign nationals already present on their territory (Caprioglio and Rigo, 2020). For instance, in some countries, instructions were given to extend work permits for migrant workers among immigrants with permits that ended in the period covered by the state of alarm. Moreover, the hiring of immigrants and the unemployed was made more flexible so that they could take these jobs (Cinco Días, 2020). During the state of alarm, the agrarian organizations sometimes opened job banks to cover the jobs that hitherto had been occupied by seasonal workers coming especially from Eastern Europe and North Africa (Heraldo, 2020).

The tourism sector has been another sector seriously affected by the crisis. After the lockdown, national governments recommended that people should spend their summer holidays inside their countries. Thus, national tourism was able to support the domestic hotel and restaurant industry facing border closures and the lack of international mobility (Riss and Sizgetvari, 2020). Between 2020 and 2021, rural tourism destinations benefited from the good booking situation, and visitor numbers were even higher than before the crisis in many rural areas (Díaz, 2021; OECD, 2020a).

One of the things most changed by the pandemic has been the way people travel. The restrictions, the sense of security offered by rural areas and the uncertainty have favoured the arrival of tourists in rural areas in Europe. Many people have discovered destinations that, without the COVID crisis, they would not have visited (Díaz, 2021).

Other measures focused on the social protection of immigrants, especially during the socioeconomic recovery in 2021. The majority of EU and OECD countries applied some exemptions from health measures for migrant workers in essential occupations and sectors to facilitate entry into the territory and rapid access to the labour market (European Commission, 2020). By that time, some regional governments (like the autonomous communities in Spain) developed social protection actions, being concerned with maintaining the living conditions of the immigrant population. According to national legislations, certain groups of immigrants were not covered by health assistance, but these governments temporarily included those people in the health system.

Moreover, during 2020 and 2021, other measures taken to promote communication and information for immigrants were also major issues for the inclusion of TCNs. In this regard, several national governments, public institutions like regions and provinces and third-sector organizations carried out actions in order to spread guidelines and information. The initiatives were diverse, such as making videos and brochures in different languages to support COVID-19 prevention and access to social services (ASGI, 2020). Furthermore, legal and health information on Corona provisions for migrants and regulations was translated into several languages (Aragón Hoy, 2020).

As a consequence of the COVID-19 crisis, in October 2021 governments such as that of Spain approved a change of the immigration regulations that will facilitate the granting of residence and work permits to minors and young foreigners who have immigrated alone to Spain (Euronews, 2021). It is a measure for young immigrants – between 16 and 23 years old – that will allow them to live and work legally in Spain, many of them employed in essential activities and in rural areas. This measure focuses on essential workers and also on preventing vulnerability among immigrants.

New opportunities for immigrants in rural and mountainous regions of Europe

Foreign immigrants within these regions have been more affected by COVID-19 pandemic than the local population. This is due to the characteristics of their jobs and to their living and working conditions: job precariousness, overcrowded housing, ghettoization or due to the lack of linguistic understanding that limited their access to information and services. For instance, many of those who were working in the tourism sector – hotels and restaurants – in personal and domestic care, suffered the closure of many companies and mobility restrictions in the territory and were fired from their jobs. Moreover, they are often at risk of becoming scapegoats in times of crisis, due to stereotypes and general mistrust within local communities.

The measures mainly developed by the public sector but also by some private and third sector institutions have helped people to cope with the new challenges, and they have supported rural regions and their inhabitants in becoming more resilient.

Presently, and from a resilience point of view, the COVID-19 crisis seems to have not only negative impacts, since it has created new opportunities for several rural and even remote regions, fostering new narratives and lifestyles appreciating alternative spatial behaviours: (a) to ensure safety distances and to manage social distancing and health security in a way radically different from that adopted in the city, all related to low population density; (b) to define new, place-based policies and new forms of local participation based on renewed regional autonomies; (c) to promote more sustainable lifestyles, reducing human environmental impact related to hypermobility and fostering the rediscovery of the role of the community, nature / the environment, and face-to-face relationships. As a consequence of these measures, rural areas can gain in attractiveness to present to residents and to newcomers like internal and international immigrants.

The COVID-19 crisis has clearly shown the importance of the foreign population residing in rural areas for the functioning of the entire economic and social system, at national as well as EU level.

Consequently, the COVID-19 crisis can accelerate new trends in regional development policymaking. It has considerably accelerated several megatrends, such as digitalization (OECD, 2021). The increase in remote working could be a game changer for the spatial equilibrium between urban and rural areas, which could have significant implications for regional development and rural policy.

References

Aragón Hoy. (2020). *El Gobierno de Aragón dará de alta a todos los migrantes en el sistema sanitario de forma temporal hasta el 30 de mayo*. Aragón Hoy. April 8. 2020. Available at: www.aragonhoy.net/index.php/mod.noticias/mem.detalle/area.1050/id.258553 (Accessed 15 December 2020).

Ardis, S. and Laczko, F. (2020). 'Introduction – Migration Policy in the Age of Immobility', *Migration Policy Practice*, 10(2), pp. 2–7.

ASGI. (2020). *Associazione per gli Studi Giuridici sull'Immigrazione*. Available at: www.asgi.it/coronavirus/ (Accessed 15 January 2021).

Baglioni, S., et al. (Eds.). (2021). *The Impact of Social and Economic Policies on Migrants in Europe* (MATILDE Deliverable 3.1 and 4.1). https://matilde-migration.eu/wp-content/uploads/2021/07/d43-10-country-reports-on-economic-impacts.pdf.

Bianchi, M., et al. (2021). *A Comparative Analysis of the Migration Phenomenon: A Cross-country Qualitative Analysis of the 10 Country Reports on Migrants' Economic Impact in the MATILDE Regions* (Deliverable 4.4). https://matilde-migration.eu/wp-content/uploads/2021/08/d44-comparative-report-on-tcns-economic-impact-and-entrepreneurship.pdf.

Caprioglio, C. and Rigo, E. (2020). 'Lavoro, politiche migratorie e sfruttamento: la condizione dei braccianti migranti in agricultura', *Diritto, Immigrazione e Cittadinanza*, 3, pp. 32–56. Available at: www.dirittoimmigrazionecittadinanza.it/archivio-saggi-commenti/saggi/fascicolo-n-3-2020-1/650-lavoro-politiche-migratorie-e-sfruttamento-la-condizione-dei-braccianti-migranti-in-agricoltura/file (Accessed 15 January 2021).

Caputo, M. L., et al. (Eds.). (2021). *10 Country Reports on Economic Impact* (MATILDE Deliverable 4.3), pp. 8–53. https://matilde-migration.eu/wp-content/uploads/2021/07/d43-10-country-reports-on-economic-impacts.pdf.

Cinco Días. (2020). 'Parados con prestación e inmigrantes podrán trabajar como temporeros para salvar las cosechas', *Cinco Días*, April 7, 2020. Available at: https://cincodias.elpais.com/cincodias/2020/04/07/economia/1586261669_724050.html (Accessed 17 October 2021).

Dettling, D. (2020). 'Die Zukunft von Stadt und Land nach Corona. Die Kommunen werden zu systemrelevanten Akteuren des Wandels. Gastkommentar', *Wiener Zeitung*, May 6, 2020. Available at: www.wienerzeitung.at/meinung/gastkommentare/2059591-Die-Zukunft-von-Stadt-und-Land-nach-Corona.html (Accessed 10 November 2021).

Díaz, D. (2021). 'Estos son los destinos rurales de España más cotizados para este verano', *The Condé Nast, Journal of Travels*, June 30, 2021. Available at: www.

traveler.es/naturaleza/articulos/destinos-rurales-pueblos-de-moda-verano-2021/21296 (Accessed 10 November 2021).

Euronews. (2021). 'Spain to Ease Restrictions on Young Migrants to Legalise Status', *Euronews*, October 10, 2021. Available at: www.euronews.com/2021/10/19/spain-to-ease-restrictions-on-young-migrants-to-legalise-status (Accessed 3 January 2022).

European Commission. (2020). *Maintaining Labour Migration in Essential Sectors in Times of Pandemic. Series of EMN-OECD Informs on the Impact of COVID-19 in the Migration Area, European Migration Network.* Available at: https://ec.europa.eu/homeaffairs/sites/homeaffairs/files/docs/pages/00_eu_inform3_labour_migration_2020_en.pdf (Accessed 9 November 2021).

Gruber, M., et al. (2021). 'The Impact of the COVID-19 Pandemic on Remote and Rural Regions of Europe: Foreign Immigration and Local Development', *GSSI Discussion Paper Series in Regional Science & Economic Geography*. Discussion paper No. 2021–15, November 2021, 38 pp. Available at: www.gssi.it/images/DPRSEG_2021-15%20(1).pdf (Accessed 4 January 2022).

Guadagno, L. (2020). 'Migrants and the COVID-19 Pandemic: An Initial Analysis', *Migration Research Series*, Nr. 60. Geneva: International Organization for Migration.

Heraldo. (2020). 'Avalancha de solicitudes para trabajar en el campo, de las que el 90% están presentadas por inmigrantes sin papeles', *Heraldo de Aragón*, April 16, 2020. Available at: www.heraldo.es/noticias/aragon/2020/04/16/avalancha-de-solicitudes-para-trabajar-en-el-campo-de-las-que-el-90-estan-presentadas-por-inmigrantes-sin-papeles-1369855.html (Accessed 12 November 2021).

Laine, J. (ed.). (2021). *10 Country Reports on Qualitative Impacts of TCNs* (MATILDE Deliverable 3.3), April 2021. https://matilde-migration.eu/wp-content/uploads/2021/06/D33-10-Reports-on-QualitativeImpacts.pdf.

Landis MacKellar, F. (2020). 'COVID-19: Demography, Economics, Migration and the Way Forward', *Migration Policy Practice*, 10(2), pp. 8–14.

Liem, A., et al. (2020). 'The Neglected Health of International Migrant Workers in the COVID-19 Epidemic', *The Lancet*, 7(4), p. e20. http://doi.org/10.1016/S2215-0366(20)30076-6.

Membretti, A. (2020). 'Compulsion to Proximity? Mobility, Proximity and the Role of Rural and Mountain Areas after the COVID-19 Crisis', *International Conference "Bodies in the Climate Change Era"*. Institute of Body & Culture, University of Konkuk, Seoul (South Korea), May 29–30, 2020.

OECD. (2020a). *Tourism Policy Responses to the Coronavirus (COVID-19)*. Available at: www.oecd.org/coronavirus/policy-responses/tourism-policy-responses-to-the-coronavirus-covid-19-6466aa20/ (Accessed 11 November 2021).

OECD. (2020b). 'COVID-19 and Key Workers: What Role Do Migrants Play in Your Region?', *OECD Policy Responses to Coronavirus (COVID-19)*, November 26, 2020. Available at: www.oecd.org/coronavirus/policy-responses/covid-19-and-key-workers-what-role-do-migrants-play-in-your-region-42847cb9/.

OECD. (2021). 'Territorial Impact of COVID-19: Managing the Crisis and Recovery across Levels of Government', *OECD Policy Responses to Coronavirus (COVID-19)*, May 10, 2021. Available at: www.oecd.org/coronavirus/policy-responses/the-territorial-impact-of-covid-19-managing-the-crisis-and-recovery-across-levels-of-government-a2c6abaf/.

Papademetriou, D. G. and Hooper, K. (2020). 'Commentary: How is COVID-19 Reshaping Labour Migration?', *International Migration*, 58(4), pp. 1–4.

Riss, K. and Sizgetvari, A. (2020). 'Reisewarnungen ausgeweitet, Urlaub in Österreich empfohlen', *DerStandard.at*, April 8, 2020. Available at: www.derstandard.at/story/2000116649989/reisewarnungen-ausgeweitet-urlaub-in-oesterreich-empfohlen (Accessed 20 December 2020).

3

Three comments to the Manifesto

3.1 Clearing the "smoky skies"

Fabrizio Barca

The time has come to "imagine another world," unless, as Arundhati Roy (2020) fears with reference to the postpandemic phase, we want to tackle migration, climate, viruses and authoritarian challenges by "dragging the carcasses of our prejudice and hatred…our dead ideas, dead rivers and smoky skies behind us." Rural and mountain areas – let us call them *inner areas*, as we do in Italy, to underline their distinguishing feature, which is remoteness – and their relation to urban areas can and must be at the centre of such radical rethinking. The MATILDE Project makes a long-awaited step in this direction by considering immigrants as a broad category of "people on the move" in search of a better way of life, and by connecting this phenomenon to the unused potential of *inner areas* and the failure to recognize the rights of their inhabitants. Rather than addressing the anger of rural people with compassionate money transfers and tackling migration as an emergency while ignoring migrants' rights and values, MATILDE provides a solid basis on which to address the aspirations and rights of both human groups by using the same policy approach.

In the ten theses that MATILDE offers the public, there is not a hint of the rhetoric or "hard-core cosmopolitanism" whereby the understanding of other people's culture comes at the cost of disregarding one's own (Appiah, 2006). The theses are drawn from the encounter between highly qualified academic research and the evidence furnished by innovative experiences and movements throughout Europe. They represent a call to the authorities at both national and EU level to pay greater attention to their societies, to take a deep breath, sweep away all the wrong tools and "smoky skies" of the last decades, and reframe "remoteness" and "migration" as a value for Europe by starting from existing experiences. The "smoky skies" which have for decades obscured the concern for people living in inner areas are immediately apparent. They are not the result of fate but of culture and policy action. Let me lay a few facts on the table by drawing on my paper

DOI: 10.4324/9781003260486-15

entitled "Place-based Policy and Politics" (Barca, 2019). Three policy approaches have dominated the scene for far too long.

First, *space-blind institutional reforms*. This approach starts from the widely accepted premise that institutions matter for development. But it then proceeds by arguing that institutional changes can and should be entrusted to technocrats and experts at national or supranational level, designing and recommending / enforcing one-size-fits-all "best practice." The following assumptions are implicitly or explicitly made: a single "institutional model" independent of contexts exists; the technocrats know what that model is; it can be implemented by writing complete contracts, i.e., where the possibility exists to predict all possible contingencies for the future; local elites are either benevolent in implementing recommendations or they can be made to behave in compliance with the contracts; finally, the role of citizens consists mostly in exiting from places if institutions are not adequate, thereby creating the incentive for their reform. No opportunity is given for the people living in a specific place to voice their knowledge and their preferences during the process of policy design. This approach has repeatedly failed to tackle the problems of "left-behind" areas because its assumptions are wrong: context matters for the effectiveness of institutions; technocrats have limited knowledge; contracts are inherently incomplete; local elites are often unwilling to innovate since they derive power and income from a condition of local backwardness; most citizens do not have the means either to assess whether exiting is good for them or to actually do so. While failing to design reforms and investments suited to places, this policy has also produced resentment because of its systematic and deliberate failure to consider local people's aspirations and knowledge.

These negative effects have been compounded by a second approach: *corporation-led agglomerations*. This approach starts from the correct observation that agglomeration – the concentration in cities of business activity and living – is a primary driver of growth, and large global firms are "machines" for collecting and recombining knowledge. However, this approach makes two unwarranted assumptions: that the negative externalities of agglomerations can eventually be balanced and brought under control – they cannot – and that corporations take decisions that balance the interests of all stakeholders, an assumption made even more unreasonable by the dominance of a "shareholder-value" culture. Based on these assumptions, the approach calls for policy action, i.e., major public investments, to accommodate the agglomeration choices made by corporations – *a de facto* spatial policymaking led by them – and to promote people's mobility. Whilst this approach contributes to the growth of cities and to the fortunes of many people within cities, it is also responsible for major inequalities within and among cities and has actively contributed to the rural/urban divide by not taking into account the potential and aspirations of people in noncity areas.

Faced with the territorial inequalities produced by these two approaches, governments have repeatedly adopted a third approach aimed at *"compensating" for those inequalities,* mostly in peripheries, deindustrializing towns and rural areas. The objective has been to avoid – or, as it has turned out, to postpone – the social consequences of the two main policies while obstinately pursuing their implementation. While sustaining employment and income in the short term, compassionate compensations have produced perverse effects by turning local elites into intermediaries of public funds and rentiers, weakening people's motivation for mobilization and change and impinging on their dignity and self-respect.

People living in *inner areas* have been the first victims of these three interconnected approaches, which have led to a deterioration of fundamental services, the designing of rules (at both national and EU level), that persistently ignore people's specific features and aspirations, a failure to understand their opportunities and an improper relation between urban and rural areas and between mountain and metropolitan towns. Major recognition inequalities have accompanied economic and social injustice: people of *inner areas* have perceived that neither their values and norms (sameness) nor their needs, aspirations and role as guardians of the land, environment and *paysage*, and as labourers engaged in agriculture, forestry and pasture (oneness), were being recognized. When new migrations took place, when migrants or refugees were suddenly bussed into empty village buildings or abandoned hotels, or when inadequate services were opened to them, the local people felt that both their sameness and oneness were being mortified – further mortified – by diversity. Thus fuelled was an authoritarian dynamic which called for tough authorities sanctioning "deviant behaviours."

This does not have to be the case. Experiences and studies throughout Europe show that an alternative exists, one which can address the rights and the aspirations of both local inhabitants of inner areas and of migrants. MATILDE provides us with new lenses through which to see the foundations of this alternative. The diversity of both cultural and natural resources within a short distance – the distinguishing feature of most inner areas – is an extraordinary resource with which to improve well-being, and it provides a "lateral vision" that enhances innovation. The scarcely populated space between places responds to a need for less anthropic pressure. An accelerated digital transformation enables a redesign of life-work organization whereby remote areas are no longer seen as just providing a temporary escape from "the real world." Adjusting to climate change and investing in the circular economy opens up new opportunities to many rural and mountain areas.

Wherever these new opportunities have been harnessed, we see at work a policy approach that is radically different from the past. A place-based approach whereby people in inner areas are given the power and the knowledge to expand their sustainable, substantial freedom, i.e., to overcome the obstacles to the full development of their own potential "without compromising," as Amartya Sen puts it, "the capacity of future generations to have similar, or more, freedom" (Sen, 2009). This takes place through an improvement in the access and quality of essential services and in the opportunity to innovate. Active citizens' organizations and movements, creative local leaders, engaged teachers, doctors and administrators, innovative entrepreneurs and other agents of change are always behind these experiences, as the MATILDE Project stresses and as the Forum on Inequality and Diversity in Italy keeps rediscovering. However, for these experiences to be sustainable, to produce lasting trust and not to be ruled out as "niche episodes," the active engagements of national governments and, behind them, of European institutions are necessary.

This is the innovative framework within which international migration can become part of the solution, not of the problem. From the point of view of inner areas, international migration is seen by MATILDE as part of a general flow of people on the move: young people moving in search of freedom and work while maintaining links with their origins; young people moving in so that they can apply their knowledge acquired "outside" to local idiosyncratic resources; professionals redesigning their life-work organization; seasonal work in agriculture and tourism sectors; elderly persons from the middle and upper working classes looking for a place where their savings are worth more. All these individuals are called upon to play a game-changing role which can be brought to the fore by a place-based approach. In this context, international migrants can add significant value, contributing to the renaissance of abandoned places with their skills, values, sentiments, commitment, entrepreneurship and networks.

Within a place-based framework, the inclusion of migrants can cease to be – as MATILDE stresses – a "one-way integration" and become a "nonlinear and reciprocal interaction through which new population groups negotiate new cultural meanings and concrete rights of citizenship." It can grow into a win-win situation for both old and new inhabitants. International migrants can fully express their value without confining themselves to their own inner circles, a tendency due to breed resentment in the EU-born children of the second generation. The inhabitants of inner areas, rather than being presented by well-to-do "hard-core cosmopolitan" urban elites with their duty to accommodate diversity, are given the chance to see their own social rights as the primary concern of authorities and to perceive the fulfilment of migrants' social rights as complementary to theirs.

This is the ultimate fulfilment of Kwame Anthony Appiah's idea of "partial cosmopolitanism," whereby, in a context where we feel commitment to care and closeness to any human being, it is reasonable to be "partial to those closest to us" (Appiah, 2006). A concept also expressed recently by Pope Francis in the Encyclical "Fratelli Tutti" (2020):

> Just as there can be no dialogue with "others" without a sense of our own identity, so there can be no openness between peoples except on the basis of love for one's own land, one's own people, one's own cultural roots. I cannot truly encounter another unless I stand on firm foundations, for it is on the basis of these that I can accept the gift the other brings and in turn offer an authentic gift of my own. I can welcome others who are different, and value the unique contribution they have to make, only if I am firmly rooted in my own people and culture.

Inclusion thus becomes a process whereby those furthest from us become gradually part of those closest to us, a two-way process that can only start if the original local community feels recognized.

Let us not fool ourselves. We are not there yet by any means. We have only understood from research, evaluation and experiences that this "other world" can be imagined and achieved. This is the remarkable step forward made by the MATILDE Project. But in the majority of cases around Europe, both the rights of rural and mountain communities and those of international migrants are ignored. Old-fashioned top-down, space-blind policies are still the order of the day. More than 13 years since the "Agenda for a Reformed Cohesion Policy," which, thanks to the foresight of an EU Commissioner, Danuta Hubner, proposed a radical shift of EU regional policies towards a place-based approach, the funds of the Resilience and Recovery Fund very often risk being used with no attention paid to the aspirations, knowledge and voices of many inner-area communities (Barca, 2009). And in several rural areas of Europe, international migrants are indeed being treated as slaves and live in dwellings that are unacceptable by any human standard – as I have witnessed in my own country. Even when immigrants are treated with decency or have regular contracts, they very rarely have the chance or are given the reassurance needed to speak out and truly participate in the decision-making process within their new communities.

This state of affairs makes the MATILDE Report and its ten theses even more valuable. They should be brought to the attention of national and EU authorities and become part of the advocacy activity of civil society organizations of all sorts. They should also be shared among the local communities that are venturing into new terrain, among the new ruralism and new highlander movements, as well as among every movement currently mobilizing in order to clear the "smoky skies." They can be the catalyst for a growing body of ideas and practices. They can reduce the solitude of, and provide support to, young local administrators emerging throughout Europe who are generating some hope that the existing, conservative and exhausted political elites can be renewed.

References

Appiah, A. K. (2006). *Cosmopolitanism: Ethics in a World of Strangers*. New York: W.W. Norton & Co.

Barca, F. (2009). *An Agenda for a Reformed Cohesion Policy: A Place-based Approach to Meeting European Union Challenges and Expectations*. Available at: I (europa.eu).

Barca, F. (2019). 'Place-based Policy and Politics', *Renewal a Journal of Social Democracy*, 27(1).

Encyclical. (2020). *Fratelli tutti*. Available at: Fratelli tutti (3 ottobre 2020) | Francesco (vatican.va) (Accessed 28 December 2021).

Roy, A. (2020). 'The Pandemic as a Portal', *Financial Times*, April 3.

Sen, A. K. (2009). *The Idea of Justice*. London: Allen Lane & Penguin Books.

3.2 Tying territory, society and transformation together

A Manifesto with an integral approach

Manfred Perlik

Why a Manifesto?

The EU-funded MATILDE Project was launched after the peak of the so-called refugee crisis of 2015. Besides its humanitarian focus, it examines territorial inequality and spatial justice in light of examples of mountainous areas as a kind of laboratory of peripheral living conditions.[1] In European mountain and peripheral areas, the hosting of refugees has not until now been of major public concern. It can be assumed that peripheral areas are not the places that refugees most want to live in. Nor is it an easy task for the original local population to host overnight larger number of immigrants. Therefore, the hosting of refugees in mountain areas can be considered a social innovation. In addition, the coincidence of various global crises (climate, pandemic, global value chains) superposed and strongly influenced the project, especially when considering their interdependency and mutual self-reinforcement. In this way, the MATILDE Project links three key themes together: mobility, territorial development and social innovation.

Mobility and uneven territorial development

At least since the forced development of the social division of labour, i.e., since the colonial conquest of the world, territorial development has proceeded unevenly, with certain territories either gaining or losing importance. Associated with this have been population movements of immense proportions through flight and displacement driven by both explicit violence and economic pressure. Since the onset of modern capitalism, the concentration of people in cities has repeatedly registered new peaks, but these have been interspersed by contrary processes due to political, economic and humanitarian crises and which are manifest in both political mass movements and political-economic paradigm shifts. There is a recurrent pattern: in times of rapid economic growth, here are market expansion, an increasing social division of labour and

DOI: 10.4324/9781003260486-16

the concentration of residence in urban areas. By contrast, in times of recession, crisis and subsequent social paradigm shifts, marginalized areas consolidate, but without regaining their former political and economic significance (Bairoch, 1985; Pumain, 1999; Schuler et al., 2004). By adopting the approach of evolutionary economics and its concept of "trajectory," one can explain why prosperous societies gain new development options while, at the same time, certain windows close so that there is no movement in reverse. Hence, the cities at the top of the territorial hierarchy today are those that in the past have managed to maintain urban and metropolitan growth amid unbridled global competition. This development has been enabled by a certain constancy of those cities' ruling class over time and their continuous attraction of all sorts of capital generated elsewhere, combined with a flexible alternation of investment, disinvestment and reinvestment. The MATILDE Manifesto depicts these two key elements in its chapters on inequal development, the explanation of migration and the search for social and spatial justice.

Transformative social innovation

Still missing is the third element that might explain transformative change in society: social innovation. I prefer to speak explicitly of *transformative* social innovation, i.e., innovation that triggers changes in the relationships between social actors and institutions and not just improved regional business models. Crucial for the definition of social innovation is the scale applied to decide what is really *new* and what *social* means to avoid social and greenwashing. In this respect, the benevolent reception of migrants is frequently a social innovation, as has been shown in Italy in the cases of municipalities such as Riace (Reggio Calabria), Pettinengo (Biella) and others (Perlik and Membretti, 2018): local populations connect to humanitarian experiences of the past and reject ethnic/nationalist/identitarian instrumentalization.[2] Engaging in the reception and inclusion of migrants may enhance cohesion within mountain communities and may increase regional identity to stabilize them. With its focus on the reception of refugees (Theses 4–7), MATILDE clearly distinguishes itself not only from ethnic nationalism but also from national pseudosocialist concepts of regional identity (in the literature often euphemistically termed "left" populism). MATILDE thus offers a strong counterforce against regional egoism (Davezies, 2015), individual exclusion and racism.

Socioterritorial relationship

The Manifesto therefore brings together three issues that are usually separated. Migration experts typically tend to consider the positive or negative impact of migrant reception for the benefit of the interest groups represented by them, i.e., immigrants, or in the opposite case, incumbent inhabitants. Regional developers have adopted "best practices" in promoting identity and distinction, and they seek to expand on international markets. Innovation experts hope for an entrepreneurial competitive advantage. This sectoral view obscures the causes of the current crises; it quickly favours particular interests and clientelism, and it ultimately inhibits progress in achieving the UN Sustainable Development Goals (SDGs). The Manifesto breaks up this sectoral view. It has the merit of showing in a condensed form to a broader public the interdependence among territorial disparities on a global and a regional scale, migrant flows and new options for transformation.

Critique

However, this integral approach is only partially successful, primarily because the ten theses have been written by individual authors and are thus conditioned by a heterogenous sectoral logic. The theses are additive rather than interrelated or interlocking. On the one hand, this means that part of the overall view is lost; on the other hand, it loses concreteness. This has consequences for the substantive positioning of humanitarian refugee reception and generates an overly optimistic view with regard to the development of peripheral areas:

- The message is not self-evident. There is no consensus in European societies that peripheral spaces should not be left to themselves; there is still a strong belief in market forces. The same applies to the reception of migrants: European societies are divided as never before; the debate is only mitigated by reduced migrant flows and the predominating topic of the COVID crisis. This problem cannot be remedied with a new narrative (i.e., a better communication).
- Indeed, COVID has given new functions to peripheral areas as temporary escapes from the sanitary insecurity of cities, as in the 14th century. However, it is completely unclear whether this can halt or even reverse the loss of importance that has occurred in recent years. On the contrary, there is a danger that the observable tendency towards a monofunctional, selective use of mountain areas will be reinforced.
- Therefore, although we see certain signs for transformation, the overlying euphoric stocktaking obscures the view for the missing link in the analysis of territorial disparities and the search for spatial justice.

In regard to the development of mountain regions, neoclassical economists rely on regulation by market forces, architects discover new creativity through art and aesthetics instead of production and the ecological mainstream advocates a strict separation of cultural space and biosphere, i.e., tends to abandon sparsely populated territories. The *metrophilia* mentioned in the Manifesto (a very good term!) is an indicator that the connection between the well-being of the mountain periphery and the metropolitan cores – a prerequisite for a prosperous society – has fallen out of sight (as Thesis 8 clearly states). The appeals by the community of alpine research and development for an upgrading of the periphery – in recent years constantly repeated – do not help, because it remains unclear who (which social actors) should engage in such upgrading. The problem lies deeper. The hope that an immediate and enduring trend reversal will start with the COVID crisis seems too short-sighted and premature. It is true that, as a result of the pandemic, the population figures of the big cities are currently stagnating.[3] This reflects that, if a big city can no longer exploit its structural advantages of manifold social interaction, it becomes too expensive for its citizens, and economic agglomeration advantages turn into *dis*advantages. The advocates of the free market might feel vindicated. But nobody should be deceived. With the recovery after lockdown, the city also may regain its structural advantages, not only via the greater opportunities for interaction, but also via the concentration of the built environment and infrastructures that impacts as a lock-in factor ("too big to fail"). When these cities develop problems, their weight – grown over decades – is so heavy that costly innovations are introduced first and foremost in them.[4] Conversely, we see the selective valorization of mountain areas through aesthetization and gentrification under the label "landscape," whereby an environmentally destructive infrastructure is built at the same time (for the example of the Himalayas: Jacquemet, 2018; Naitthani and Kainthola, 2015; for the Alps: Perlik, 2019). The generation of new dynamic hotspots and new peripheries is thus reproduced again and again. It therefore makes sense to look for the missing link that brings together migrant flows, spatial disparities and transformative social innovation.

Searching for the missing link

It is worth re-reading Rosa Luxemburg's seminal 1913 work *The Accumulation of Capital*, which deals with one of the fundamental contradictions of capitalist societies: the compulsion to achieve perpetual growth for perpetual accumulation and the search by capital owners for ever new ways to privatize the commons. The book begins with an in-depth critique of Marx's second volume of *The Capital*, which ideally assumed a completed penetration of market relations for 19th-century Europe. Not at all, Luxemburg says. Once the reproduction of the population has been achieved (including measures for the damage caused by environmental degradation), it is necessary to find new investment opportunities for the surplus accumulated previously. In the 19th and 20th centuries, this expansion of markets took place as colonialism through the destruction of indigenous cultures and social practices – Luxemburg describes in detail the destruction of pre-colonial peasant societies in Algeria, India and China, as well as that of the smallholder settlers in North America. Today, in postcolonialism, the global instabilities are generated by geostrategic interests driving migration flows, as well as disinvestment from industrial production sites (relocations) and individual habitats (rural-urban domestic migration) to new places: for example, with the widespread introduction of second homes and the social enforcement of multilocal living practices. In short, capitalist market penetration (and also its territorial expression, urbanization) is never complete. It is a regularly recurring "primitive accumulation" that takes place whenever established products no longer generate the necessary profit margins, but depending on political regulations. The modern expression for it is "the paradigm of permanent growth" with the fascination of the buzzword *innovation* based on Schumpeter's "creative destruction."

Why this excursus into political economy? Because the logic of creating new commodities to reinvest overaccumulated benefits may serve as the missing link to build a coherent critique for the transformation of spatial and social relations. Consequently, the brief COVID-induced trend interruption of metropolitan concentration evaporates. On the contrary, the current new functions of mountain areas become highly selective and dependent on the development of the new platform economy invented in the metropolitan regions. New commodified functions are the following: mountain retreat for reasons of personal security, the investment of value in real estate because the other investment vehicles have lost performance or the search for additional tourism models because guaranteed snow coverage and the demand for ski tourism are declining. These specializations on global leisure markets follow a logic of economic autonomy, but they reinforce the dependence on external developments; in this sense, they narrow the future options for action instead of widening them. Breaking with this liberal-productivist[5] logic, developed after the Fordist crisis of the 1980s, would once again necessitate a change of regime, i.e., a profound transformation of the conditions under which social wealth is produced (accumulation) and distributed (regulation).

Thesis 9 in the MATILDE Manifesto presents the foundational economy as an economic approach tailored to peripheral regions and the reception of migrants. However, this does not change the fact that the dominant economic processes follow a liberal-productivist logic which even the multiple crises cannot immediately put into question. But the foundational economy, together with similar approaches, like the solidarity economy and a revitalization of the cooperative movement, may contribute to creating countertendencies for a change of regime. There is a considerable research potential in the entire range of alternative economic and political models beyond the current offer-oriented regional competition.

For a new relationship between the collective and the decentral, between identity and solidarity

Although it does not seem justified to be optimistic in a short-term change in favour of the peripheries, the current cumulation of several different crises offers the political potential for more profound emancipatory changes, i.e., transformative social innovation. They can be successful in the long run if the current division of societies into privileged centres and low-value-creating peripheries can be overcome. This requires abandoning the illusion that peripheries can manage on their own only if they are particularly innovative in commercial terms and compete with each other in an offer-oriented manner. Today's split between so-called "rural" areas and the internationally oriented metropolitan areas has produced devastating distortions, of which the USA, Brazil and Eastern Germany / Europe are only the most prominent examples (many of the Manifesto theses refer to them by citing Andrés Rodríguez-Pose's Brexit analysis). Representatives of mountain areas have long insisted on the superiority of decentralized structures and on strengthening regional identities. However, the current political polarization in many European countries, based on spatial types with their different life chances, puts this recurrent mantra into perspective. The result is often a mere shift of political power in favour of other, more assertive groups of actors, grounded in nationalist-regionalist thinking which promote social exclusion and racist discrimination against even more disadvantaged people. At the same time, they do little to change the fundamental structural strength of metropolitan regions and their dominance over the peripheries.

Rather, the reverse conclusion should be drawn: if decentralized structures today favour fragmented identities and milieus in which both "city" and "countryside" feel exploited by each other, then they must indeed be fundamentally questioned. This includes the strengthening of lowland/mountain linkages so that differences in productivity are mitigated with, for example, a possible conclusion to abandon regional business models mainly based on high-end long-distance tourism.

How this transformation of an offer-oriented, identity-based competition into more solidary structures in larger territorial units could come about – especially under conditions of worldwide migration flows that will not decrease – requires a great deal of further research like that, for example, being currently conducted in the German-speaking area by the "Critical Land Research Working Group" (Maschke et al., 2020) for peripheral areas. In regard to social innovation and a solidarity economy, the CIRIEC network of the University of Liège has carried out significant groundwork. There is a multitude of such initiatives. Many of them do not know about each other. It is important to fill this lack of networking for further, cross-national migration research in mountain areas. For this, a wide spreading of this first version of the Manifesto is very desirable.

Notes

1 Many European mountain ranges are well developed, especially the Alps, but compared to metropolitan regions, they are peripheries.
2 It is assumed that hosting institutions practice honest arguments and try to find good solutions for the people involved on both sides. But we should always be aware that remote places may also be used to "hide" refugees to avoid integration and to get rid of them easily.
3 For example, school enrolments have declined in Paris, Lyon and Marseille.
4 This is the diagnosis for the current liberal-productivist regimes. Conditions may change. There are strong arguments for innovation due to peripherality (Glückler et al., 2022; Mayer et al., 2021) which may become, under changed regimes, more than a niche.
5 I prefer this term to the common "neoliberal" because, on the one hand, *neoliberal* has become a very common pejorative term, although it is not precisely defined, and on the other hand, this term does not treat the fundamental question of what is produced and under what conditions for the (animate and inanimate) environment.

References

Bairoch, P. (1985). *De Jéricho à Mexico*. Paris: Villes et économie dans l'histoire.

Davezies, L. (2015). *Le nouvel egoïsme territorial*. Paris: Seuil.

Fourny, M.-C. (2014). 'Périphérique, forcément périphérique? La montagne au prisme de l'analyse géographique de l'innovation', in Attali, M., Granet-Abisset, A.-M. and Dalmasso, A. (eds.), *Innovation en Territoire de Montagne*. Le Défi de l'Approche Interdisciplinaire. Montagne et Innovation. Grenoble: PUG.

Glückler, J., Shearmur, R. and Martinus, K. (2022). 'From Liability to Opportunity: Reconceptualizing the Role of Periphery in Innovation', *SPACES Online*, 17(2022–01). Toronto and Heidelberg. Available at: www.spacesonline.com (Accessed 1 February 2020).

Jacquemet, E. (2018). 'Réinventer le Khumbu: la société sherpa à l'ère du "Yak Donald's"', in Fourny, M.-C. (ed.), *Montagnes en mouvements. Dynamiques territoriales et innovation sociale*. Grenoble: PUG.

Luxemburg, R. (1913[1985]). 'Die Akkumulation des Kapitals', *Gesammelte Werke Bd.* 5: 5–411, Berlin: Dietz.

Maschke, L., Naumann, M. and Mießner, M. (2020). *Kritische Landforschung: Konzeptionelle Zugänge, empirische Problemlagen und politische Perspektiven.* Berlin: Rosa Luxemburg-Stiftung, Studien 1/2020; Bielefeld: Transcript.

Mayer, H., Tschumi, P., Perren, R., Seidl, I., Winiger, A. and Wirth, S. (2021). 'How Do Social Innovations Contribute to Growth-independent Territorial Development?', *Case Studies from a Swiss Mountain Region. Die Erde*, 152(4), pp. 218–231.

Naitthani, P. and Kainthola, S. (2015). 'Impact of Conservation and Development on the Vicinity of Nanda Devi National Park in the North India', *JAR-RGA*, 103(3). Available at: http://rga.revues.org/3100 (Accessed 29 June 2021).

Perlik, M. (2019). *The Spatial and Economic Transformation of Mountain Regions. Landscapes as Commodities*. Routledge Advances in Regional Economics, Science and Policy. London: Routledge.

Perlik, M. and Membretti, A. (2018). 'Migration by Necessity and by Force to Mountain Areas: An Opportunity for Social Innovation', *Mountain Research and Development*, 38(3), pp. 250–264.

Pumain, D. (1999). 'Quel rôle les petites et moyennes villes ont-elles encore à jouer dans les régions périphériques?', in Perlik, M. and Bätzing, W. (eds.), *L'avenir des villes des Alpes en Europe*. Bern: Geographica Bernensia P36, Revue de Géographie Alpine, 87(2): 167–184.

Schuler, M., Perlik, M. and Pasche, N. (2004). *Non-urbain, campagne ou périphérie – où se trouve l'espace rural aujourd'hui*? Berne: Office fédéral du développement territorial.

3.3 The need for a less territorial, more people-centred and relational approach

Annelies Zoomers

Background

After many decades in which attention has been one-sidedly concentrated on the "urban future," since 2021 closer attention is being paid to the "left-behind areas," where the remaining populations have increasingly been left to look after themselves. According to the EC's rural vision document (and based on public consultation in rural areas), a large proportion of "local people" are discontented: almost 40% of respondents said that they felt left behind by society and policymakers (in spite of having advantages linked to farming and agritourism). About 50% of respondents stated that infrastructure was the most pressing issue for rural areas, and 43% said that access to basic services and amenities, such as water and electricity, as well as banks and post offices, was an urgent requirement. Around 93% believed that the attractiveness of rural areas depends on the availability of digital connectivity, 94% said that it depends on basic services and e-services and 92% stated that it rests on improving the climate and the environmental performance of farming. Due to limited connectivity, underdeveloped infrastructure and a lack of diverse employment, rural areas are less desirable for younger people to live in (EU, 2021).

The MATILDE Manifesto and the EU's long-term vision coincide in their plea for closer attention to reinvestment in rural "left behind" areas, which are home to 137 million people (almost 30% of the EU's population). In calling for the "renaissance of remote places," MATILDE highlights the need for local autonomy – and also the need to move away from neoliberal development in the direction of alternative development while being explicit about the positive role of immigration and newcomers. Remoteness is presented as a "strength" because it makes it possible to rethink "business as usual" and take the "local" as the new point of departure for rebuilding society. This is different from the EU's long-term approach, where much emphasis is given to "improving connectivity both in terms of transport and digital access" and making these areas contribute to "green growth." The top priorities are

DOI: 10.4324/9781003260486-17

making rural areas more prosperous by diversifying economic activities and improving the value added of farming and agri-food activities and agritourism and enhancing resilience by "preserving natural resources and greening farming activities to counter climate change while also ensuring social resilience through offering access to training courses and diverse quality job opportunities" (EU, 2021). The goal is to foster economic, social and territorial cohesion and "respond to the aspirations of rural communities" (which is less strong than the Manifesto's call for local autonomy), but community empowerment is acknowledged to be important for successful interventions.

There are a number of dilemmas and risks which need to be resolved before rural remote areas are flooded by projects coming from the outside to contribute to sustainable, cohesive and integrated "green" development. What are the core challenges and how can it be ensured that "local" people benefit?

Challenge 1: *Remoteness versus connectivity: how to prevent resource-grabbing, displacement and gentrification*

MATILDE's call for the "renaissance of remote areas" comes at a time when large numbers of projects are awaiting implementation for "green growth" and/or achievement of climate goals. Given that there are currently billions available to be spent on the "green transition" and climate change, how to prevent the projectification of landscapes and outsiders taking over, and what will be the long-term for possibilities to achieve integrated, cohesive and inclusive development? What are described as "remote" areas with potential for alternative development in the MATILDE Manifesto can easily be seen as empty areas by policymakers and investors, providing cheap land suitable for green investments. The "green deal lobby" is extremely powerful, and investors are eager to find places in which to install large-scale windmill and solar parks or acquire large tracts of forest land for ecotourism or inexpensive land to be given back to nature (space for reforestation or as flood zones). Since COVID-19, urban residents are increasingly interested in buying second homes, raising prices for land and real estate, there is a risk that local people are not powerful enough to counterbalance futures that are projected by outsiders (Zoomers et al., 2021).

Assessing the possible consequences, it might be interesting to draw a parallel with the global "land rush" – the rapid increase in large-scale land acquisitions following the 2007–2008 world food price crisis as the consequence of large-scale investments in land for food and biofuels, but also with tourism complexes, hydro dams, infrastructure, nature conservation, etc. (Borras and Franco, 2014; Cotula, 2012, 2014; Deininger and Byerlee, 2011; Zoomers, 2010; Kaag and Zoomers, 2014). It is apparent that local populations are often not well-informed or powerful enough to play a real role in decision-making, and landscapes have experienced rapid transformations, restricting people's access to open commons (land, water, forests, etc.). Investments generate new employment opportunities, but jobs are often given to outsiders (with better educations) or are poorly paid, temporary jobs. Local groups are penalized by their loss of access to commons, or displacement and compensation is often not enough to buy new land, due to rapidly increasing prices. This leads to gentrification and pushes people towards more marginal, low-cost areas and makes them more vulnerable to climate risks of flooding and drought. It is now commonly acknowledged that large-scale land investments have, in many places, been at the cost of local livelihoods and local landscapes, leading to enclosures, displacement and resettlement of vulnerable groups and the fragmentation of landscapes. Given the lack of an underlying masterplan, the global land rush has, in many places, led to the "projectification" of landscapes, which has limited local people's manoeuvring space (Zoomers, 2010). Large-scale land investments in plantations for biofuels, mining, dams, solar and windmill parks, etc. have resulted in landscape destruction – loss of biodiversity and deforestation – and the exclusion of local populations (the rapid growth of "no go" areas).

In conclusion, the inflow of projects will not automatically generate positive results. Given the EU plans to improve connectivity – both in transport and digitally – this is supposed to go hand in hand with the creation of new employment opportunities. But before implementing any "new rural vision," it is important to protect the rights of existing people (including the provision of compensation arrangements) and carefully reflect on what investments are required, taking communities' priorities into account and considering the entire range of intended and unintended consequences, also in the long run. According to the EU's new vision, a necessary requirement for the socio-economic enhancement of rural areas is to improve their accessibility (to make them more attractive), but this may go hand in hand with rising land prices and gentrification (see also Thesis 7), pushing vulnerable groups aside (Zoomers et al., 2016a, 2016b). In other words, paying closer attention to "remote areas" and putting the "local community" at centre stage in decision-taking requires time and good preparation so that communities can take an active role in attracting the right type of investments, anticipate the intended and unintended consequences and find ways to benefit from profit-sharing.

Challenge 2: *Local communities are not homogenous – how to deal with diversity and how to define community-based development?*

In the MATILDE Manifesto, as well as in the rural vision document, much emphasis is given to local communities – giving them a key role as agents of change. Whereas the MATILDE Manifesto stresses the need to base "local development" on the foundational economy, the European Commission seems to focus on the potential benefits of green growth. Local development impacts are mainly described in terms of income and employment generation, while rural areas are described as target areas. "Local" is conceived as spatially bound and small, and local "communities" are seen as "homogeneous, territorially fixed, small and homogeneous wholes with shared norms" (Agrawal and Gibson, 1999, p. 633), able and willing to make desirable collective decisions such as when negotiating with investors. The challenge is how to deal with diversity: local communities, even if they existed as homogeneous "wholes" – which is, in reality, never the case – are increasingly fragmented due to differential impacts of influences from the outside as well as differences in the abilities of diverse locals to link to nonlocal opportunities. What does "local development" mean, and how can investment plans be brought into alignment with local people's priorities?

Given that remote areas host various groups with usually different needs and aspirations, the question is how to make sure that plans are compatible with (local) views on "local development" and how to contribute to improved levels of well-being. To conceptualize "development" and gain a better understanding of rural dynamics (going "deeper" than "local community"), it is useful to employ two interlinked concepts influential in development studies: (i) the notion of "development as freedom" advocated by Amartya Sen (1999) and his "capability approach," for which the basic concern of human development is "*our capability to lead the kind of lives we have reason to value*," and (ii) the Sustainable Livelihood Approach (SLA), which captures how people build their livelihood using different kinds of capital (Bebbington, 1999; De Haan and Zoomers, 2005; Kaag et al., 2004; Zoomers et al., 2016A).

Typical of the livelihood approach is that – in contrast to the earlier tendency to conceive poor people as passive victims – it highlights the active, and even proactive, role played by the (rural) poor. The emphasis is on seeing people as agents actively shaping their own future, focusing not on what poor people lack but rather on what they have (their capital) and on their capability (Sen, 1999; de Haan and Zoomers, 2005). Given this reality, opportunities for (people in) remote rural areas to take the lead in defining their own future will depend on whether they are able to build consensus on what "development as freedom" is about and make strategic use of the various capitals.

More than elsewhere, people in remote, resource-poor areas are often obliged to combine a range of strategies in order simply to survive; individuals may engage in multiple activities, and the various members of a household may live and work in different places or opt for a development path characterized by multitasking and income diversification. There is a tendency towards livelihood diversification, i.e., "a process by which… households construct an increasingly diverse portfolio of activities and assets in order to survive and to improve their standard of living" (Ellis, 2000, p. 15). In many cases, the bulk of the income of the rural poor no longer originates from agriculture; people have multiple income sources. Distinctions between rural and urban livelihoods are increasingly difficult to make: on the one hand, rural people (who are formally registered as "villagers") live and work in the city for most of the year (staying with family members and working in construction, housekeeping, etc.); the better-off in the rural areas buy parcels in the cities in order to be able to give their children a better education (exit strategies). On the other hand, people in the urban sphere start growing food crops in the cities (new types of urban agriculture), whereas urban elites are increasingly expanding into the rural sphere by becoming the owners of rural land. Affluent urbanites obtain the land by foreclosing on loans. Increasing land values have led them to look upon land as an attractive commodity for investment purposes.

In addition to multitasking and the blurring of the rural/urban interface, there is a trend in which rural people increasingly develop multilocal livelihoods. Rapid urbanization and the improvement of communications and transport technology have significantly increased mobility. Growing numbers of rural poor now engage in urban and rural life, commuting from the countryside to urban centres on a daily basis, sometimes travelling large distances to earn additional money as temporary migrants, and also, international migration is rapidly increasing. Considerable numbers of rural poor are no longer rooted in one place; although they maintain relations with their home communities, they are also attached to other places and function in larger networks.

Challenge 3: *People do not live in containers – the importance of linkages and corridors*

In endeavouring to achieve spatial justice, targeting remote, deprived areas – and providing them with projects – will not work. It is important to look outside: people in remote areas do not live in containers, and inspection of their capitals and capabilities evidences that they have geographically dispersed networks – even the smallest groups. Part of the potential for new ways forward arises from the inside, from the locally available resources which could become the basis for new developments, but the dynamics will in the longer run largely come from the outside. Understanding the broader spatial networks – and positionality of remote areas – is a *sine qua non* for understanding the potential dynamics, even in the most isolated places. We argue that in "rethinking Europe," it is time for a "mobilities turn" (Sheller and Urry, 2006) that challenges the sedentarist assumptions often still in the minds of policymakers and practitioners trying to plan for the future (Manifesto, thesis 7). Discussions on how to stimulate local development usually end up by calling for actions within fixed and confined settings ("the project area"), but globalization connects even distant people and places (Zoomers and van Westen, 2011). Rather than depending essentially on "local resources," livelihood opportunities are increasingly shaped by positionality and the way people are attached to and participate in translocal and transnational networks.

In conclusion

We very much support the MATILDE Manifesto's plea to put remote rural areas back on the policy agenda while stressing the need for community-based development and contributing to social and spatial justice. Since the summer of 2021, these areas have also been targeted by the EU as focal points for implementing "green growth" strategies – and billions of euros are ready to be spent on the "green deal" and projects related to climate change (see Thesis 1). EU recovery plans offer the occasion to put remote areas at the centre of the debate on the future of Europe. This is positive, but it also has risks. Citing the global land rush, we have shown the danger that remote ("empty") areas may be invaded by large-scale investment projects from the top down (e.g., solar and wind parks, ecotourism projects and/or biofuel plantations, etc.) which do not offer space for "new rural and mountain narratives" (Thesis 2) and without really taking root in the territory concerned.

In order to achieve MATILDE's goal of turning remote areas into breeding grounds for alternative development and social innovation, priority should be given to strengthening the self-determination capacity of local communities. Communities are, however, not homogeneous, and community empowerment is required in order for priorities to be set (and consensus to be built) in regard to the desired pathways of change. Given the characteristics of these areas (i.e., the high incidence of inflow and outflows of people), community building should be seen as a moving target. The "renaissance" of remote areas requires an open approach: rather than defining a "local" community on the basis of belonging to a particular territorially bounded space, it is important to include the people moving in and out of it. Given current realities – especially in rural and remote areas – part of the population will constantly cross boundaries (and be regularly outside) but can still make important contributions to the flourishing of the so-called "foundational" economy.

We agree with the MATILDE Manifesto (Thesis 9) that we need new criteria for a community belonging based on people's actual contribution to economic, social, cultural and political life, instead of one based on only legal and normative assumptions of belonging. Hence, newcomers are part of community as long as they perform "acts of citizenship" and take an active role in the provision, defence and reproduction of the local commons. Adopting this stance means that we need to distance ourselves from pre-established ideas about "integration" and "assimilation."

Finally, the dynamics and development potential of remote areas will usually not depend on local factors. Rural and mountainous areas cannot be seen as stand-alone places (Manifesto Thesis 3). They form part of wider networks, and positionality is one of the major determinants for being able to attract the human and financial resources required for alternative development or not. We argue that instead of focusing on the (socially constructed) confined space in which local people live, more attention should be paid to the relational aspects of livelihood and development, acknowledging that there is a need (even urgency) to deal also with transformations coming from the outside. Instead of trying to keep people in place and focusing on local assets, the challenge is to have a trustful and productive relationship with the outside. Establishing an extended network with people in different localities will help to mobilize resources in multiple directions, getting the best from various worlds. A less territorial, more people-oriented and relational approach could help to achieve a more sustainable and inclusive society.

References

Agrawal, A. and Gibson, C. C. (1999). 'Enchantment and Disenchantment: The Role of Community in Natural Resource Conservation', *World Development*, 27(4), pp. 629–649.

Bebbington, A. (1999). 'Capitals and Capabilities: A Framework for Analysing Peasant Viability, Rural Livelihoods and Poverty', *World Development*, 27(12), pp. 2021–2044.

Borras, S. and Franco, J. C. (2014). *Towards Understanding the Politics of Flex Crops and Commodities. Implications for Research and Policy Advocacy*. Transnational Institute. Think Piece Series on Flex Crops and Commodities.

Cotula, L. (2012). 'The International Political Economy of the Global Land Rush: A Critical Appraisal of Trends, Scale, Geography and Drivers', *The Journal of Peasant Studies*, 39(3–4), pp. 649–680.

Cotula, L., et al. (2014). 'Testing Claims about Large Scale Land Deals in Africa: Findings from a Multi-country Study', *The Journal of Development Studies*, 50(7), pp. 903–925. http://doi.org/10.1080/00220388.2014.901501.

De Haan, L. and Zoomers, A. (2005). 'Exploring the Frontier of Livelihood Research', *Development and Change*, 36(1), pp. 27–47.

Deininger, K. and Byerlee, D. (2011). *Rising Global Interest in Farmland. Can It Yield Sustainable and Equitable Benefits?* Washington, DC: The World Bank.

Ellis, F. (1998). Household Strategies for Rural Livelihood Diversification. *Journal of Development Studies*, 35(1), pp. 1–38.

EU. (2021). *A Long-term Vision for the EU's Rural Areas*. Available at: https://ec.europa.eu/info/strategy/priorities-2019-2024/new-push-european-democracy/long-term-vision-rural-areas_en (Accessed 10 November 2022).

Kaag, M., van Berkel, R., Brons, J., de Bruijn, M., van Dijk, H., de Haan, L., Nooteboom, G. and Zoomers, A. (2004). *Poverty is Bad: Ways Forward in Livelihood Research.* Interuniversitary Research School for Resource Studies for Development CERES. The Hague: CERES.

Kaag, M. and Zoomers, A. (eds.). (2014). *The Global Land Grab. Beyond the Hype*. London: Zed Books, pp. 1–16, ISBN: 978-1-78032-894-2.

Sen, A. (1999). *Development as Freedom*. New York: Alfred A. Knopf.

Sheller, M. and Urry, J. (2006). 'The New Mobilities Paradigm', *Environment and Planning*, 38, pp. 207–226. http://doi.org/10.1068/a37268.

Zoomers, A. (2010). 'Globalization and the Foreignization of Space: The Seven Processes Driving the Current Global Land Grab', *Journal of Peasant Studies*, 37(2), pp. 429–447.

Zoomers, A., Leung, M., Otsuki, K. and van Westen, A. (2021). *Handbook of Translocal Development and Global Mobilities*. Elgar Publishers. Available at: www.e-elgar.com/shop/gbp/handbook-of-translocal-development-and-global-mobilities-9781788117418.html.

Zoomers, A., Leung, M. and van Westen, G. (2016b). 'Local Development in the Context of Global Migration and the Global Land Rush: The Need for a Conceptual Update', *Geography Compass*, 10(2), pp. 56–66. http://doi.org/10.1111/gec3.12258. Available at: http://onlinelibrary.wiley.com/doi/10.1111/gec3.12258/pdf.

Zoomers, A. and van Westen, A. (2011). 'Introduction: Translocal Development, Development Corridors and Development Chains', *International Development Planning Review*, 33(4), pp. 377–388.

4
Conclusion

Reconstruction of remoteness as a new centrality and dialogical cocreation of living together

Anna Krasteva, Andrea Membretti and Thomas Dax

"There is a place for everybody and everybody should be in their place," states a well-known saying. Remote and rural regions have been taught to know their place (Ching and Creed, 1997). This Manifesto challenges this social, political and representational *status quo* and engages in a theoretical reconstruction of remoteness as a new centrality. Manifesto is a specific genre which mixes analytical insights with strategic visions of what is and what should be a pathos for change and confidence in change-makers.

The conclusion structures the arguments and messages of the Manifesto in regard to remoteness and to migration in remote, mountain and rural regions. The "Symbolic battles for the core of Europe" section examines the people/places nexus in the context of the interplay between policies and politics or how the remoteness of mountain and rural areas is de- or reconstructed by European, national, regional and local policies on the one hand, and local actors on the other (see Theses 1, 2, 3, 8, 9 and 10.). "The dialogical cocreation of living together" section structures the impact of immigration along the axes from integration to innovation and from governance to citizenship (see Theses 4, 5, 6 and 7).

While the thrust of the Manifesto is synthesized within these two lines of argumentation, it should be underlined that all theses should be seen as a joint manifestation of altered conceptual positions, new emphasis on emotional foundations and reappraisal of intrinsic potential of remote spaces and people, attributed through their diverse aspects in the various theses. The conclusion is an elaboration of the insights of all authors of the Manifesto. Their theoretical, empirical and policy-oriented ideas and contributions are gratefully acknowledged here.

DOI: 10.4324/9781003260486-19

Symbolic battles for the core of Europe

From remotization to sense-making policies for territorial justice and translocal solidarity

The notion of political remoteness is dominated by the image of a powerful centre that defines places outside and distant as peripheral and marginalized, the spatial fringe interfering with the political deficits. This imaginary constructs remoteness as downgraded and deprived of self-government. The marginalization of remoteness leads to "remotization" (Membretti, 2021) – the increasing physical and symbolic distance between and within rural/mountain and urban areas and their populations. This Janus-faced concept encapsulates the ambivalent process of reciprocal (cultural and physical) removal and sociospatial rarefaction. This process is accompanied by a widespread perception of unprecedented remoteness, a widening of everyday living space, and a stretching/weakening of connections/ties that can lead both to social resentment/isolation and to new opportunities for local development, innovation and new lifestyles (see Theses 1, 2, 3, 8, 9 and 10).

The Manifesto pleads for the restoration of places and spaces to people so that territorial equity and translocal solidarity become the core of "next-generation Europe." Remote places are reconceptualized as:

- People's vital and multifaceted world of experience, resisting neoliberal globalized homogenization, grounding their future on the diversity of their cultural and positional resources, values and potential.
- The basis for place-sensitive and place-based policies capturing the option of physical distance as a benefit and appreciating the space in between that characterizes scarcely populated areas.

The Manifesto is an urgent call for a new and different public voice, a "lateral vision" rich with potential to overcome weak social capital and community development and respectful of a wider range of themes and spaces otherwise dominated by the logic of "central places" and agglomeration.

Reterritorialization – from being crisis-driven to being driven by values and social innovation

COVID-19 is an incentive for change in the symbolic battle for attractiveness between remote and urban places (see Thesis 10). Digital nomads enjoy combining the advantage of "global" work with the charm of choosing by oneself where to live. The local becomes more globalized; the global becomes more individually localized. The more detached and distant,

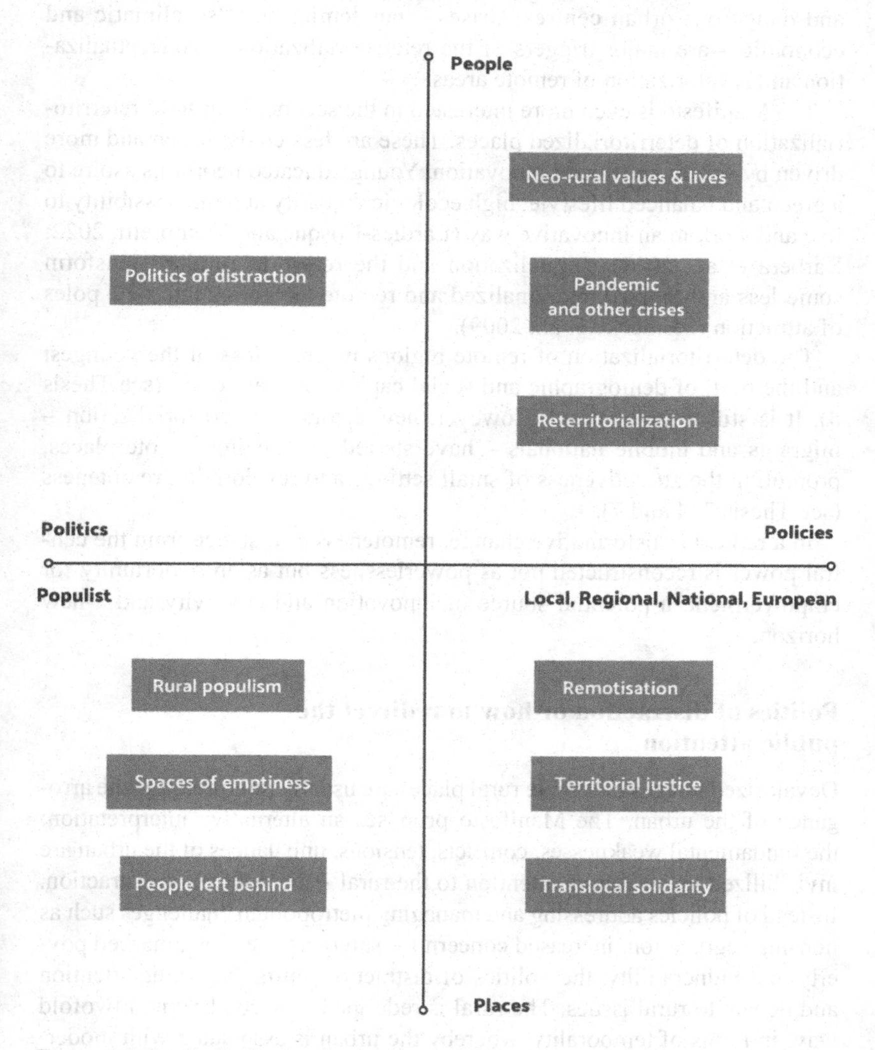

Figure 4.1 Symbolic Battles for the Core of Europe

Source: Elaboration by the author

the more attractive: this is the new normality in pandemic times. Europe is experiencing the unprecedented momentum of remoteness and remote places. While local populations can still critically and negatively conceive their places as cut off from essential services for health crisis management, newcomers positively evaluate remoteness as an escape from overcrowded and dangerous urban centres. Crises – pandemic, but also climatic and economic – are major triggers of the reterritorialization, reconceptualization and revalorization of remote areas.

The Manifesto is even more interested in the second, long-term reterritorialization of deterritorialized places. These are less crisis-driven and more driven by values and social innovation. Young educated neorurals aspire to a green and balanced lifestyle, high ecological quality and the possibility to live and work in an innovative way (Lardies-Bosque and Membretti, 2022; Barbera et al., 2018). Digitalization and the return to rurality transform some less anthropized, marginalized and remote territories into new poles of attraction (Steinecke et al., 2009).

The deterritorialization of remote regions means a loss of the youngest and the best, of demographic and social capital, of brain drain (see Thesis 8). It is still a major trend. However, new actors of reterritorialization – migrants and mobile nationals – have started revitalizing remote places, promoting the attractiveness of small settings and revalorizing remoteness (see Thesis 2, 4 and 7).

In a radical transformative change, remoteness as distance from the central power is reconstructed not as powerlessness but as an opportunity for empowerment, a potential source of innovation and creativity and a new horizon.

Politics of distraction or how to redirect the public attention

Devalorized images of remote rural places are usually interpreted as the arrogance of the urban. The Manifesto proposes an alternative interpretation: the fundamental weaknesses, conflicts, tensions, unbalances of the urban are invisibilized by switching attention to the rural – the politics of distraction. Instead of policies addressing and managing metropolitan challenges such as housing segregation, increased concern for safety and security, marked poverty and vulnerability, the politics of distraction shifts the public attention and debate to rural issues. The rural is redefined as secondary in a twofold way: in terms of temporality, whereby the urban is associated with modernity, development, enlightenment and the rural with tradition, identity and roots; in terms of politics, whereby far-right populism is exported to rural regions more sensitive to their identity politics because of their more homogeneous ethnocultural fabric. Placing populism in rural areas and emphasizing its connection to specific rural features may be understood as a way for political power-holders to shift the focus from nearby urban shortcomings to unwanted processes in more remote, rural places (see Thesis 8).

For countering these distorted images, the Manifesto proposes active forms of citizenship that revalorize the neglected remote regions (see Thesis 8).

Spaces of emptiness, socioeconomic, spatial and nostalgic deprivation and populism in remote places

There is an intense drain on social, cultural and economic capital in remote rural areas which creates "spaces of emptiness." Populism is not a pathological matter; it is the consequence of the social-economic and political disparities that have haunted Europe in the past few decades, an expression of socioeconomic, spatial and nostalgic deprivation. In times of crisis, individuals tend to establish communities in order to protect themselves and to cope with uncertainty, insecurity, unemployment, exclusion and poverty in the age of deindustrialization. This need becomes even more impelling in remote places, where it takes the form of an appeal to homogeneity, the past, heritage, culture, and religion. The growing affiliation of the supporters of right-wing populist parties in remote places with culture, nativism, authenticity, religiosity, traditions, myths and civilizational rhetoric provides them with an opportunity to establish solidarity networks against structural problems.

The Manifesto sends a threefold message. The European Commission should recognize that the EU's "unity in diversity" motto does not successfully translate into the lives of lower-educated, geographically immobile and socioeconomically and spatially deprived social groups, which tend to see both "diversity" and "unity" as challenges to be tackled. Policymakers should interpret right-wing populism as the consequence of long-lasting social-economic and political inequalities rather than as a pathological matter. Mainstream political parties should focus on social-economic and psychological issues to communicate better with their electorates, who are likely to feel socioeconomically, spatially and nostalgically deprived (see Thesis 8).

Dialogical cocreation of living together

European policies for resilient rural areas and the Manifesto's contribution

"Rural areas are the fabric of our society and the heartbeat of our economy." Thus, **Ursula von der Leyen**, President of the European Commission, persuasively launched the long-term vision for stronger, connected, resilient and prosperous rural areas.[1]

The Manifesto makes a fourfold contribution to the new strategic vision by:

- Framing, formulating and promoting a new narrative on rural and mountain areas, from places left behind to places of social innovation, transformative social change and attractive lifestyles.
- Providing a bottom-up perspective on active, solidarist and creative citizenship, participatory action research and local engagement.
- Enhancing the new trend in European policymaking for policies to be based on citizens' values and identities, and thereby complement evidence-informed policies.
- Expanding local agency with an inclusive approach that involves also newcomers, in-migrants, international migrants who contribute to creating the place and shared life worlds.

Participatory action development and self/assessment for empowering local stakeholders

The local level, where governance meets the "vital worlds," is the primary one for creative sense-making policies and innovative practices. The Manifesto promotes policy tools and practices that can bring together citizens, migrants, local authorities and participatory research to pursue the goals of connecting, collaborating and creating. They help to maximize the contribution of diverse local actors to tailor-made practices of integration and social cohesion.

"Policies need to take into account and reflect the values and identities of citizens," a recently published report states (Sefkovic, 2021). Values are core drivers of change. A better understanding of diverse values and multiple identities enables policymakers to design more resilient, meaningful and inclusive policies (ibid.). The Manifesto promotes this new trend in European policymaking by endorsing participatory policy tools (see Thesis 5).

One of these tools is the participatory evaluation of policies and practices (see Thesis 5). Its roots are in Kurt Lewin's participatory action research theory. Lewin sought to "raise the self-esteem of minority groups" (Adelman, 1993, p. 7) that foster their "independence, equality, and co-operation" (Adelman, 1993, p. 7). Evaluation processes should be participatory, democratic and interactive so as to reveal blind spots and include different stakeholder perspectives (Racino, 1999). The MATILDE Project implements various participatory evaluation methods (Kordel et al., 2021), such as peer-to-peer exchange via local case study working

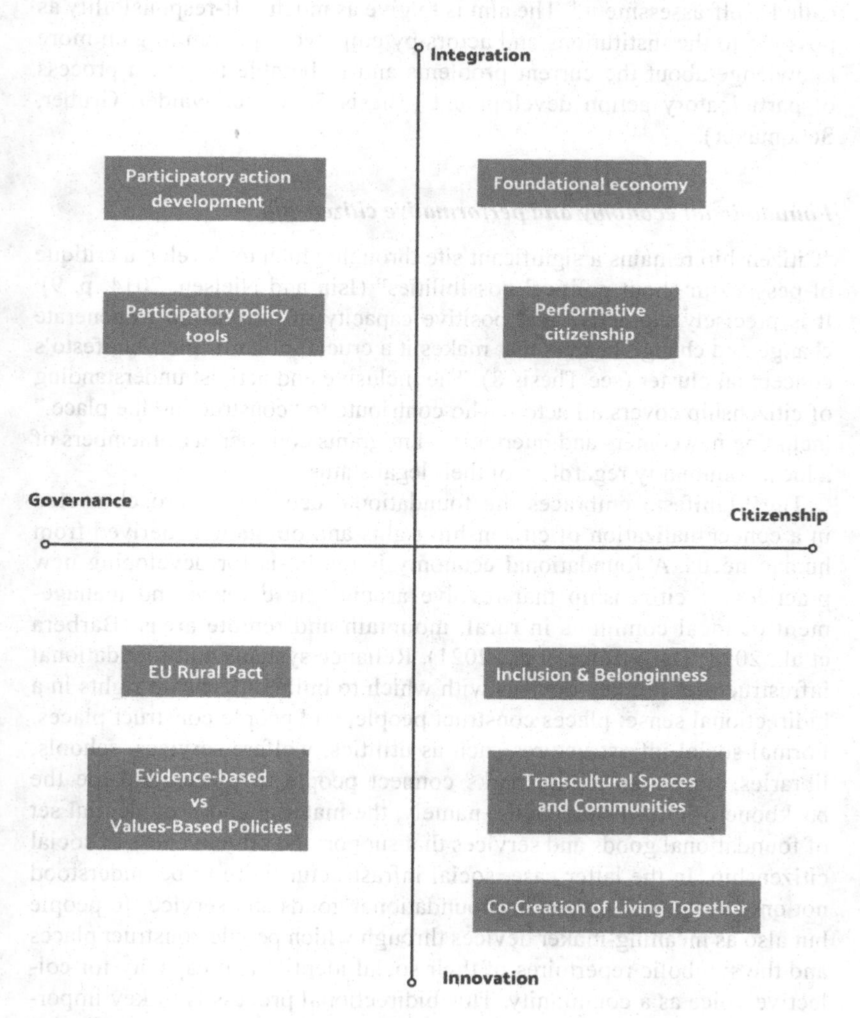

Figure 4.2 Dialogical cocreation of living together

Source: Elaboration by the author

groups and (policy) round tables, or coevaluation approaches using participatory action research methods conducted via productive interaction between scholars and local partners. The MATILDE Practitioner Toolbox fosters the peer-to-peer approach and enables practitioners to assess and analyse the situation in their community by themselves. This activity is called "self-assessment." The aim is to give as much self-responsibility as possible to the institutions and actors by empowering them to gain more knowledge about the current problems and to be able to start a process of participatory action development (Thesis 5, Aigner-Walder, Gruber, Schomaker).

Foundational economy and performative citizenship

"Citizenship remains a significant site through which to develop a critique of pessimism about political possibilities" (Isin and Nielsen, 2014, p. 9). It is precisely the active and positive capacity of citizenship to generate change and change-bearers that makes it a crucial pillar of the Manifesto's conceptual cluster (see Thesis 8). The inclusive and activist understanding of citizenship covers all actors who contribute to "constructing the place," including newcomers and international migrants considered as members of a local community regardless of their legal status.

The Manifesto embraces the foundational economy approach rooted in a conceptualization of citizenship rights and obligations derived from human needs. A foundational economy is the basis for developing new practices of citizenship that revolve around the defence and management of local commons in rural, mountain and remote areas (Barbera et al., 2018; Dalla Torre et al., 2021). Reliance systems and foundational infrastructures are key devices with which to build citizenship rights in a bidirectional sense: places construct people, and people construct places. Formal social infrastructures such as utilities, welfare services, schools, libraries, hospitals, public parks connect people to places and are the backbone of citizenship rights: namely, the material and providential set of foundational goods and services that support the effectiveness of social citizenship. In the latter case, social infrastructures are to be understood not only as "pipes" conveying foundational goods and services to people but also as meaning-maker devices through which people construct places and the symbolic repertoires of their social identity and capacity for collective voice as a community. This bidirectional process is of key importance for place-based policies and organized experiments that support citizenship in rural, mountain and remote places. Newcomers are considered as part of the community when they perform "acts of citizenship," that is, when through their labour, civic engagement, cultural sharing and appropriation, they take an active role in the provision, defence and reproduction of the material conditions that feed citizenship and basic needs (see Thesis 9).

Citizenship rights, foundational economy and spatial justice are the minimum building blocks of a place-based planning perspective on citizenship rights and the dynamics of newcomers' inclusion in rural, mountain and remote areas (see Thesis 9).

Innovating and cocreating the living together

The rural and remote regions of Europe offer great potential not only for immigrants themselves and their respective host societies but, above all, also for something new to be created together and shared with others in the process itself of redefining "we." The inclusion of migrants is a nonlinear and reciprocal process in which both parties must not only take an active part but also be prepared to change in order for the social boundaries to become effectively blurred. A new understanding of being local, of belonging, needs to be sought through processes of inclusion and mutual recognition. These require continuous negotiation, but they fuel a social innovation in which the focus should be shifted from integration and assimilation to the cocreation of new transcultural spaces, economies, and communities. Migrants can have a remarkable social and also economic impact long before they are considered to be properly integrated. That is to say, impact does not automatically necessitate integration; integration does not automatically imply impact. The pertinent policies could yield greater benefits for all with a more pragmatic focus on the incorporation of migrants into the local social life and culture. It is time to centre the conversation on inclusion, engagement and belonginess rather than the traditional concept of integration, which gives migrants the responsibility to "integrate" into local society and often punishes those who fail to do so, thus wasting valuable means to reinvigorate rural and remote regions for the sake of normative ideals which have been shown to obstruct rather than facilitate inclusion. This approach promotes social inclusion as a nonlinear and reciprocal interaction through which new population groups negotiate new cultural meanings and concrete rights of citizenship with the existing populations within systems of socioeconomic, legal, and cultural relations whose basic characteristics need to be considered if a sustainable, equitable and resilient society is to be created for all. The role played by existing migrant communities in welcoming and easing the inclusion of newcomers should be more effectively utilized in the process. The resulting communities will not only be different but will also be better adapted to thriving in the context of the current era's seemingly endless uncertainty. Through such a broad positive impact, the prevailing – often reserved – attitudes towards migration can be improved, and a virtuous circle created whereby migrants are likely to be considered not as burdens but as valuable resources for local development (see Thesis 6).

Note

1 https://ec.europa.eu/commission/presscorner/detail/en/IP_21_3162.

References

Adelman, C. (1993). 'Kurt Lewin and the Origins of Action Research', *Educational Action Research*, 1(1), pp. 7–24. http://doi.org/10.1080/0965079930010102.

Barbera, F., Negri, N. and Salento, A. (2018). 'From Individual Choice to Collective Voice. Foundational Economy, Local Commons and Citizenship', *Rassegna Italiana di Sociologia*, LIX(2), pp. 371–397.

Ching, B. and Creed, G. (1997). *Knowing Your Place: Rural Identity and Cultural Hierarchy*. New York: Routledge.

Dalla Torre, C., Gretter, A., Membretti, A., Omizzolo, A. and Ravazzoli, E. (2021). 'Questioning Mountain Rural Commons in Changing Alpine Regions. An Exploratory Study in Trentino, Italy. Available from: https://www.researchgate.net/publication/351456554_Questioning_Mountain_Rural_Commons_in_Changing_Alpine_Regions_An_Exploratory_Study_in_Trentino_Italy [accessed May 14, 2022].

Isin, E. and Nielsen, G. (eds.). (2014). 'Introduction: Globalizing Citizenship Studies', in Isin, E. and Nyers, P. (eds.), *Routledge Handbook of Global Citizenship Studies*. London and New York: Routledge, pp. 1–11.

Kordel, S., Sauerbrey, D., Spenger, D., Dalla Torre, C. and Weidinger, T. (2021). Methodological Framework: MATILDE Toolbox (Draft 1.7.21, MATILDE Deliverable 2.7). https://mail.google.com/mail/u/0/?tab=rm&ogbl#inbox/KtbxLzFvSGlGKQLlBBBxsgRlJXjzhrLHkL?projector=1&messagePartId=0.1.

Lardies-Bosque, R. and Membretti, A. (2022, forthcoming). 'In-migration to European Mountain Regions: A Challenge for Local Resilience and Sustainable Development', in Schneiderbauer, S., Szarzynski, J. and Shroder, J. (eds.), *Safeguarding Mountains – A Global Challenge. Facing Emerging Risks, Adapting to Changing Environments and Building Transformative Resilience in Mountain Regions Worldwide*. Amsterdam: Elsevier.

Membretti, A. (2021). 'Remote Places of Europe and the New Value of Remoteness', *MATILDE: Migration Impact Assessment to Enhance Integration and Local Development in European Rural and Mountain Areas*, September 2021. http://doi.org/10.13140/RG.2.2.15779.78886.

Racino, J. A. (1999). 'Qualitative Evaluation and Research: Toward Community Support to All', in Racino, J. A. (ed.), *Policy, Program Evaluation, and Research in Disability: Community Support for All*. New York, London and Oxford: Haworth Press, pp. 3–22.

Sefkovic, M. (2021). 'Foreword', in *Values and Identities. A Policymakers' Guide*. EC. Brussels: Joint Research Centre.

Steinecke, E., Čede, P. and Flie, U. (2009). 'Development Patterns of Rural Depopulation Areas. Demographic Impacts of Amenity Migration on Italian Peripheral Regions', *Mitteilungen der Osterreichischen Geographischen Gesellschaft*, pp. 195–214.

Index

Note: Page numbers in *italics* indicate a figure and page numbers in **bold** indicate a table on the corresponding page.

Printed in the United States
by Baker & Taylor Publisher Services

Printed in the United States
by Baker & Taylor Publisher Services